Publics and the City

RGS-IBG Book Series

The *Royal Geographical Society (with the Institute of British Geographers) Book Series* provides a forum for scholarly monographs and edited collections of academic papers at the leading edge of research in human and physical geography. The volumes are intended to make significant contributions to the field in which they lie, and to be written in a manner accessible to the wider community of academic geographers. Some volumes will disseminate current geographical research reported at conferences or sessions convened by Research Groups of the Society. Some will be edited or authored by scholars from beyond the UK. All are designed to have an international readership and to both reflect and stimulate the best current research within geography.

The books will stand out in terms of:
- the quality of research
- their contribution to their research field
- their likelihood to stimulate other research
- being scholarly but accessible.

For series guides go to www.blackwellpublishing.com/pdf/rgsibg.pdf

Published

Geomorphology of Upland Peat
Martin Evans and Jeff Warburton

Spaces of Colonialism
Stephen Legg

People/States/Territories
Rhys Jones

Publics and the City
Kurt Iveson

After the Three Italies: Wealth, Inequality and Industrial Change
Mick Dunford and Lidia Greco

Putting Workfare in Place
Peter Sunley, Ron Martin and Corinne Nativel

Domicile and Diaspora
Alison Blunt

Geographies and Moralities
Edited by Roger Lee and David M. Smith

Military Geographies
Rachel Woodward

A New Deal for Transport?
Edited by Iain Docherty and Jon Shaw

Geographies of British Modernity
Edited by David Gilbert, David Matless and Brian Short

Lost Geographies of Power
John Allen

Globalizing South China
Carolyn L. Cartier

Geomorphological Processes and Landscape Change: Britain in the Last 1000 Years
Edited by David L. Higgitt and E. Mark Lee

Forthcoming

Politicizing Consumption: Making the Global Self in an Unequal World
Clive Barnett, Nick Clarke, Paul Cloke and Alice Malpass

Living Through Decline: Surviving in the Places of the Post-Industrial Economy
Huw Beynon and Ray Hudson

Swept up Lives? Re-envisaging 'the Homeless City'
Paul Cloke, Sarah Johnsen and Jon May

Badlands of the Republic: Space, Politics and Urban Policy
Mustafa Dikeç

Climate and Society in Colonial Mexico: A Study in Vulnerability
Georgina H. Endfield

Resistance, Space and Political Identities
David Featherstone

Complex Locations: Women's Geographical Work and the Canon 1850–1970
Avril Maddrell

Driving Spaces
Peter Merriman

Geochemical Sediments and Landscapes
Edited by David J. Nash and Sue J. McLaren

Mental Health and Social Space: Towards Inclusionary Geographies?
Hester Parr

Domesticating Neo-Liberalism: Social Exclusion and Spaces of Economic Practice in Post Socialism
Adrian Smith, Alison Stenning, Alena Rochovská and Dariusz Świątek

Value Chain Struggles: Compliance and Defiance in the Plantation Districts of South India
Jeffrey Neilson and Bill Pritchard

Publics and the City

Kurt Iveson

Blackwell Publishing

BLACKWELL PUBLISHING
350 Main Street, Malden, MA 02148-5020, USA
9600 Garsington Road, Oxford OX4 2DQ, UK
550 Swanston Street, Carlton, Victoria 3053, Australia

First published 2007 by Blackwell Publishing Ltd

1   2007

Library of Congress Cataloging-in-Publication Data

Iveson, Kurt.
    Publics and the city / Kurt Iveson.
      p. cm. — (RGS-IBG book series)
    Includes bibliographical references and index.
    ISBN-13: 978-1-4051-2732-5 (hardcover: alk. paper)
    ISBN-10: 1-4051-2732-5 (hardcover: alk. paper)
    ISBN-13: 978-1-4051-2730-1 (pbk.: alk. paper)
    ISBN-10: 1-4051-2730-9 (pbk.: alk. paper)
1. City planning—Australia—Case studies.   2. Public spaces—Australia–Case studies.   3.
Urban policy—Australia—Case studies.   4. City and town life—Australia—Case studies.   I.
Title.

HT169.A8I84 2007
307.1′2160994—dc22

                                                                        2006025006

A catalogue record for this title is available from the British Library.

Set in 10 on 12 pt Plantin
by SNP Best-set Typesetter Ltd, Hong Kong

For further information on
Blackwell Publishing, visit our website:
www.blackwellpublishing.com

For Nancy and Benji

Contents

Illustrations

Series Editors' Preface

RGS-IBG Book Series

Like its fellow RGS-IBG publications, *Area*, the *Geographical Journal and Transactions*, the Series only publishes work of the highest quality from across the broad disciplinary spectrum of geography. It publishes distinctive new developments in human and physical geography, with a strong emphasis on theoretically-informed and empirically-strong texts. Reflecting the vibrant and diverse theoretical and empirical agendas that characterize the contemporary discipline, contributions inform, challenge and stimulate the reader. Overall, the Book Series seeks to promote scholarly publications that leave an intellectual mark and that change the way readers think about particular issues, methods or theories.

Kevin Ward (University of Manchester, UK) and Joanna Bullard
(Loughborough University, UK)
RGS-IBG Book Series Editors

Acknowledgements

First, my gratitude to all the activists and policy workers who gave generously of their time and resources. This book would not have been possible without the information and insights they provided, and it is much better for the questions they asked of me.

The research and writing of this book have been conducted from home bases in four different cities, where I've benefited from the generosity and support of colleagues, friends and family. The following lovely people stimulated my thinking, assisted my writing and offered practical support: at the Australian National University, Tim Bonyhady, Brendan Gleeson, Heather Grant, Alastair Greig, Robyn Lui, Marian Sawer and Patrick Troy; at the University of Durham, Ash Amin, Kay Anderson, Peter Atkins, Donna Marie Brown, Mike Crang, Paul Harrison, Adam Holden, Emma Mawdsley, Niamh McEllheron, Gordon MacLeod, Charles O'Hara and Joe Painter; in Sheffield, Elizabeth Gagen, Paula Meth, Mitch Rose, Samuel Vardy and Glyn Williams; at the University of Sydney, John Connell, Phil Hirsch, Phil McManus, Mel Neave, Bill Pritchard and Kath Sund. Others who have provided valuable advice and support at various times include Craig Calhoun, Allan Cochrane, Michael Edwards, Ruth Fincher, Chris Gibson, the Griffiths family, Nick Henry, Jane Jacobs, Pauline McGuirk, Mark Peel, Frank Stilwell, Kylie Valentine, Kevin Ward and Sophie Watson.

Some folks' support and love have crossed continents and been with me all the way. Sean Scalmer is a great friend and colleague, sharing ideas, passions and distractions, averting crises of confidence, and just generally being in my corner. I particularly want to thank Jim, Marg and Mike Iveson for the sacrifices they made to get me to and through university in the first place, and for the inspirational examples they've set in following their own passions and commitments.

Jacqueline Scott and Angela Cohen at Blackwell have been incredibly helpful, and I also thank them for their monumental patience in dealing with a first-time author who has a loose concept of deadlines (it's finished, honest!).

Finally, this book is dedicated to Nancy Griffiths and Benji Iveson. Nancy has made huge intellectual, practical contributions to the writing of this book and her love and sense of adventure have made the years during which it was written so much fun. And thanks to little Benji. His uncle Mike gave him a t-shirt which says 'I'm small, but I know stuff'. It's true – life is already better for the stuff he's teaching us.

Publisher's Acknowledgement

The editor and publisher gratefully acknowledge the permission granted to reproduce the copyright material in this book. Every effort has been made to trace copyright holders and to obtain their permission for the use of copyright material. The publisher apologises for any errors or omissions in the above list and would be grateful if notified of any corrections that should be incorporated in future reprints or editions of this book.

Chapter One

The Problem with Public Space

Finding an Audience

In 1994 William 'Upski' Wimsatt, a hip hop columnist, graffiti-writer, and self-described college dropout from Chicago, self-published a book called *Bomb the Suburbs*. Upski conceived of *Bomb the Suburbs* as 'a book for people who don't usually read'. He especially wanted his book to be read by young people in the inner cities of the United States – a group he argued were severely disenfranchised by the growth of suburbs and the 'suburban mentality' which he set about attacking. To make sure his book would be accessible to his target audience, Upski gave his manuscript to around 50 different readers, including one 13-year-old girl at risk of dropping out of school, who was instructed to 'cross out the boring bits'.

Upski wrote *Bomb the Suburbs* after encountering difficulties trying to publish a newsletter called *Subway and Elevated*. This newsletter, produced by Upski and some friends, sought to raise awareness of the problems with cities in the United States and to discuss strategies for their renewal. They taped the newsletter to walls and train lines. This method of distribution had a message:

> The ultimate goal of *Subway and Elevated* was to revive public places in America – and call attention to their necessity – by placing works of beauty and value there that were impossible to obtain in stores. It was our little way of turning the tables on the reward structure in American life. If you drove a car, lived in the suburbs, and sent your kids to private school, then for once in your life you couldn't have one (Wimsatt 2000 (1999): 16).

This distribution method ran into some serious problems, not the least of which was its legality. In fact, Upski and the others involved were arrested for vandalism. At this point, they thought a book might provide a solution: 'No one could arrest us for a book, we thought' (Wimsatt 2000 (1999): 16).

But once the book was written, Upski confronted a new set of problems – if you want to write a book for people who don't usually read books, exactly how do you find this audience? He read parts of his book to commuters on subway platforms. He put up posters and graffiti with the title of the book, first around his home town of Chicago, and then further afield as he freight-hopped and hitch-hiked his way around America. He did what he could to get the book on the shelves of stores in Chicago and other cities – not just bookstores, but also music and clothing retailers and other places where his target audience were likely to shop. People sold Upski's book in their schools and junior colleges. He attempted to gain publicity for his book by staging a 'Bet with America'. This bet was a kind of radical alternative to Newt Gingrich's 'Contract with America', in which Upski bet that he could hitch-hike around America and walk the streets of its most feared neighbourhoods without actually getting hurt – thus demonstrating (hopefully) that the fear of strangers which seemed to characterize mainstream culture in the United States was misplaced, and that the negative hype about ghettoes served only to erase the humanity of the people who live there. He used the proceeds from sales of the book to help set up a writers' workshop with a twist:

> In my version, the group would be based not on what we wrote, but on *where* we wrote – not in cafés, bookstores or addresses on the World Wide Web, but in public places of the city (Wimsatt 2000 (1999): 18, original emphasis).

For Upski, then, 'the medium was the message'. He sought to reinvigorate the very public spaces which were under attack through the writing, reading, promotion and sales of *Bomb the Suburbs*.

Upski continued to encounter difficulties in circulating his ideas through a book. His guerrilla advertising methods and readings continued to attract the attention of urban authorities. Once again, Upski was arrested, this time during a street-corner reading, on a 'string of goofy charges' (Wimsatt 2000 (1999): 19). The publication of a book also brought Upski into conflict with a group of 'publishing industry motherfuckers' whose practices made it difficult to find places where the book could be sold – powerful publishers seemed to dictate what books were on sale and display in most shops, and the aggressive business strategies of chain retailers made life difficult for independent booksellers and other 'mom-and-pop' shops which might stock the book in neighbourhoods across the United States (Wimsatt 2000 (1999): 31–2).

Nonetheless, Upski's perseverance paid off. In 1998, a second edition of *Bomb the Suburbs* was printed by Soft Skull Press, an independent publishing house based in New York City. The new edition found its way onto shelves in independent bookstores across the United States, and was even sold by some major retail chains like Tower Records. By the time Upski's next book came out in 1999, *Bomb the Suburbs* had sold about 23,000 copies. It has continued to sell well since, with sales recently passing 40,000. Through Upski's various efforts, his book found an audience.

As Upski's experience with *Bomb the Suburbs* illustrates, 'finding an audience' is hard work. Indeed, this phrase hardly captures the difficulties of public address. Upski's audience was not simply there waiting for him to 'find' it. Rather, to 'find' an audience is to *make a public*. It is to construct a scene through which ideas, claims, expressions and the objects through which they are articulated can circulate to others. And as Michael Warner (2002: 12) notes:

> when people address publics, they engage in struggles – at varying levels of salience to consciousness, from calculated tactic to mute cognitive noise – over the conditions that bring them together as a public.

My aim in *Publics and the City* is to develop and apply a framework for understanding the urban dimensions of these struggles. How are cities put to work by those engaged in efforts to circulate ideas and claims to others, and how do their efforts in turn (re-)shape cities?

Public Address and 'Public Space'

So, how might we begin to investigate the urban dimensions of struggles over the making of publics? Existing frameworks for such investigations frequently associate the city's contribution to public-making with the existence of *public spaces* where people can (re)present themselves before an audience of strangers. But across these analyses, there are some important differences in how public space is conceptualized. One of the key differences in understandings of public space relates to the geographical dimensions of the concept. Here, we can distinguish between two dominant approaches to the concept of public space. The term 'public space' is often used to denote a particular kind of place in the city, such that one could colour public spaces on a map – this is a *topographical* approach. By contrast, however, the term 'public space' is sometimes used to refer to any space which is put to use at a given time for collective action and debate – this is a *procedural* approach. I now want to explore these important differences in some depth, in order to assess the usefulness of the concept of 'public space' for investigating the urban dimensions of public-making.

Public address and 'public space': *topographical* approaches

'Public space' is most commonly defined in a topographical sense, to refer to particular places in the city that are (or should be) open to members of 'the public'. Here, we are talking about places such as streets, footpaths, parks, squares and the like. For many urban activists and scholars, access to such places is said to be vital for opportunities both to address a/the public and to be addressed as part of a/the public. Among those who make this connection between public-making and public spaces, there is a wide-spread concern that public spaces in contemporary cities are becoming more exclusionary, and hence less accessible to those seeking to put them to work in circulating ideas and claims to others.

Upski's street-corner readings to passers-by could be considered as one example of how an urban public space (understood topographically) can be put to work for public address. Based in part on his own experiences, Upski worried that the possibilities for street-corner readings seemed to be receding in contemporary cities. He is certainly not alone in articulating a fear that urban public spaces are becoming less accessible in cities all over the world. There now exists a large (and still expanding) literature which focuses on a range of developments which are said to be making public spaces less public. Some worry about the widespread proliferation of enclosed shopping malls in the second half of the twentieth century, initially in North American cities but now well beyond. In the mall, the gathering of strangers is organized to facilitate shopping rather than speech-making. If Upski tried to read excerpts of *Bomb the Suburbs* to passers-by in a mall, he would most likely find himself escorted off the premises by the mall's private security guards (unless he'd been invited to conduct a book reading by a retailer such as Tower Records, of course). If he read on suburban street corners, Upski might be lucky to encounter any passers-by at all – if suburbia's critics are to be believed, suburbanites are more likely to be speeding in their cars between their homes and some other enclosed space in the urban archipelago, leaving public spaces deserted. Yet other analysts worry that some street corners are simply off-limits to non-residents like Upski, accessible only to residents and their invited guests ensconced behind walls, with gated entries restricting access. Even on the more densely populated and accessible street corners that still exist, an Upski reading session might be watched closely by security agencies via closed-circuit television surveillance cameras. He could find himself accused of performing without a permit and asked to move on by police charged with enforcing local ordinances designed to enhance 'quality of life' (indeed, this did happen!). Should a crowd gather to listen to Upski, they too might be asked to move on if they threaten to block the free movement of pedestrian traffic. If they dared step off the footpath and onto the road itself, thus blocking motorized traffic, this would likely provoke an even more forceful response.[1] Of course, the very fact that Upski continued to engage in street-corner readings seems to confirm the notion that the intentions of regulators are never fully realized. Upski, and many others, have fought for the right to access 'public spaces' (Mitchell 2003).

This picture is complicated by the fact that many of the very policies and technologies just described as exclusionary are often supported on the grounds that they enhance, rather than reduce, access to public space. Politicians enact measures to restore 'order' and 'quality of life' in public space on behalf of a public that they claim is intimidated by begging, threatened by graffiti, menaced by boisterous groups of teenagers, disgusted by the smell of urine or faeces they associate with rough sleepers, and inconvenienced by unauthorized political gatherings which block traffic. Here, it is argued that exclusion from public spaces is the *product of* so-called 'antisocial' and criminal behaviour. Planners and law enforcement agencies charged with the responsibility of improving public space argue that the exclusion of a troublesome minority will make public space more accessible to the well-behaved majority. And the more people use public space, the more attractive it becomes – not only to other residents, but to people from elsewhere who might be tempted to visit or relocate (see for example Carr, Francis et al. 1992). Such policy agendas are by no means the sole preserve of the political right. Similar objectives have been pursued by a range of urban administrations, from the conservative Mayor Giuliani in New York to the socialist Mayor Maragall in Barcelona.[2] For their part, retail and residential developers argue that the spaces they produce are profitable and popular precisely because they offer users the kind of shelter from the weather and/or the strangers who threaten them that is not provided in more traditional forms of public space.

These debates reflect a range of normative perspectives on what makes for good public space. In my own earlier effort to conceptualize public space, I was particularly concerned to tease out the differences in how contributors to the public space debates understood the 'public' in 'public space' (Iveson 1998). While these differences are of course important, it now also seems to me that the different positions staked out in these debates have more in common than it might first appear. Indeed, most writing that conceptualizes public space topographically shares two problematic features.

First, many arguments on behalf of better 'public space' are articulated through narratives of loss and reclamation. Where public space has been found to have become more exclusionary, it is argued that the priority for action is to reclaim it from those who are trying to capture it for their own particular purposes. The villains and the heroes change depending on who is telling the story. Some claim that public space is under threat from the actions of corporations and developers more concerned with profit than public use, while others identify the culprits as overzealous law enforcement agencies and urban authorities who value order over democratic expression, or modernist planners who value rationality over community. Yet others argue that drug dealers, teenage gangs and other 'anti-social' groups have appropriated public space through violence and intimidation. Activist groups have sprung up to 'reclaim the night' from masculine violence against women, and to 'reclaim the streets' from the automobile. Each of these narratives is concerned with the apparent erosion of public space by the actions of those who are said to be anti-public. Perhaps the twentieth century has been witness to 'the fall of public man' (Sennett 1978), and perhaps the twenty-first century threatens to bring with it 'the end of public space' (Sorkin 1992)? As Bruce Robbins (1993: viii) noted over a decade ago:

> The list of writings that announce the decline, degradation, crisis, or extinction of the public is long and steadily expanding. Publicness, we are told again and again and again, is a quality that we once had but have now lost, and that we must somehow retrieve.[3]

When concerns about exclusion are articulated through narratives of loss, they imply that public spaces used to be more inclusionary – more 'public' – before their contemporary degradation. Robbins (1993: viii) is right to complain that:

> the appearance of the public in these historical narratives is something of a conjuring trick. For whom was the city once more public than now?

The publicness that we are supposed to have lost is in fact a 'phantom', never actually realized in history but haunting our frameworks for understanding the present. Far too often, it is ambiguous and under-theorized, featuring as an afterthought to tales of exclusion and loss. Boddy's (1992: 152) rousing call for a return to 'real' public space at the conclusion of his analysis of the 'analogous city' of overhead and underground pedestrian thoroughfares is illustrative:

> A zone of coexistence, of dialogue, of friction, even, is necessary to a vital urban order; either we must return to the streets, or the analogous city must

become more like the real city and the real streets from whence it came . . .
Where the analogous city has been built, we need to find ways of opening it
up to a complete and representative citizenry – even to those who threaten,
avow causes, or cannot or choose not to consume.

Of course, not all who are concerned with the accessibility of 'public
spaces' such as streets and parks build their critiques of exclusion through
such narratives of loss. Some contributors to the public space debates have
urged against nostalgia for times and places that were by no means perfect.
Instead of idealizing past public spaces and cities, writers like Don Mitchell
have sought instead to show how access to public space is always a product
of political struggle. Mitchell realizes that public spaces have never been
'open to all' – nonetheless, the very *ideal* of a public space which is 'open
to all' circulates to powerful effect. For him, the ongoing circulation of this
ideal becomes a 'rallying point for successive waves of political activity' as
excluded groups seek inclusion in the public spaces of the city (Mitchell
1995: 117). These struggles for inclusion have:

> reinforced the normative ideals incorporated in notions of public spheres and
> public spaces. By calling on the rhetoric of inclusion and interaction that the
> public sphere and public space are meant to represent, excluded groups
> have been able to argue for their *rights* as part of the active public. And each
> (partially) successful struggle for inclusion in 'the public' conveys to other
> marginalized groups the importance of the ideal as a point of political
> struggle (Mitchell 1995: 117, original emphasis).

From this perspective, the struggle for democratic urban public space is 'an
activity involving creation and construction, not repair and retrieval'
(Phillips 1992: 50).

But even where narratives of loss and reclamation are rejected, most
topographical approaches to 'public space' share a second problematic
proposition about the relationship between public address and the city. In
essence, this shared proposition can be summarized as follows: public
address requires inclusionary and accessible urban public spaces where
people can take their place as part of the public. Indeed, Don Mitchell's
(2003) book *The Right to the City* provides one of the clearest statements
of this proposition. As Mitchell puts it, while 'the work of citizenship
requires a multitude of spaces', the 'public spaces' of the city are 'decisive,
for it is here that desires and needs of individuals and groups can be *seen*'
(Mitchell 2003: 33, original emphasis). For him, to be *part of* the public is
to be seen *in* public. Thus, to be part of the public is to have established
the right to occupy what he calls 'material' public spaces – by which he
means streets, squares, parks, and the like – and to put them to work in
acts of public representation or address. As he puts it (2003: 131), 'public

space is the space of the public'. From this quite topographical perspective, the *problem with public space* is one of 'increasing alienation of people
from the possibilities of unmediated social interaction and increasing
control by powerful economic and social actors over the production and
use of space' (2003: 140).

The proposition about the relationship between public address and the
city which is commonly held by topographical approaches to public space
has some fundamental problems. Most significantly, much of the writing
on the accessibility of urban public space is premised on a flawed con-
ceptualization of the relationship between three distinct dimensions of
publicness:

- publicness as a *context for action* ('urban public space');
- publicness as a *kind of action* ('public address'); and
- publicness as *a collective actor* ('a/the public').

When analysts propose that the urban dimension of struggles over the
conditions of public address revolve around struggles over access to
topographically defined public spaces, they (implicitly or explicitly) tend to
assume an equivalence between these three dimensions of publicness.
However, if we begin to unpack the connections between these dimensions
of publicness, we see that this equivalence simply does not hold. Most
importantly, access to a place generally considered to be 'public space' in
a topographical sense can be shown to have no fixed or privileged relationship to acts of 'public address' or to one's status as a member of 'the
public'.

When the urban dimension of public address is equated with the provision of accessible 'public spaces', this focuses our attention on a narrow
range of places which somehow qualify as 'public' rather than 'private'.
Topographical approaches miss the messy and dynamic urban geographies
of publicness. For instance, one could certainly address a public (or indeed
be addressed as part of a public) by appearing in a space conventionally
understood as 'public', such as a street, a square or a park. But one could
also address/be addressed as part of a public through action in a place conventionally understood as 'private', such as a bedroom in a domestic house
which contains a radio and a telephone. From here, one could conceivably
be *heard* (if not *seen*) as a participant in a talkback radio debate.[4] Equally,
forms of address conventionally understood to be 'private' (such as
conversations between intimate acquaintances) can take place in spaces
conventionally referred to as 'public'. Clearly, then, the conventional
understanding of the public/private distinction as it is applied to *contexts
for action* (i.e. the distinction between public and private places) is not
wholly defined with reference to the *kind of action* that takes place in these

spaces (see also Staeheli 1996). Nor can topographical distinctions between 'public space' and 'private space' be easily defined with reference to 'the public' as a *collective*. One might certainly be *seen* by strangers in a street, a shopping mall or a workplace, but such places have quite different relationships to 'the public' by virtue of the different proprietary and regulatory arrangements through which they are established and managed.[5]

Precisely because these equivalences do not hold, topographically defined concepts of 'public space' are inherently unstable. If we enquire as to the 'publicness' of any given place with reference to any of the three dimensions of publicness I have identified above, we will likely end up revealing incongruities rather than equivalence. As Michael Warner (2002: 27, 30) has argued:

> Public and private are not always simple enough that one could code them on a map with different colors – pink for private and blue for public . . . [M]ost things are private in one sense and public in another.

This applies to streets and homes as much as it applies to other things. And of course, it also applies to places that are more obviously hybrids of public and private, such as shopping malls and public toilets. Any 'where' can potentially be the context for combinations of both 'public' and 'private' action. Just as feminist theorists have argued that 'public and private do not easily correspond to institutional spheres, such as work versus family, or state versus economy' (Young 1990: 121), so too it can be argued that public and private do not easily correspond to urban places such as street versus home, or park versus shopping mall. To echo Warner, such places can indeed be 'private in one sense and public in another'. It should not be surprising, then, that even the best of the interventions in the 'public space' debates surveyed above struggle to make this slippery concept workable.

Public address and 'public space': *procedural* approaches

The incongruities raised by the different dimensions in which a space might be 'public' cannot simply be resolved if only we could find a clear definition of what constitutes a 'public' or 'private' place – such clarity would inevitably come at the cost of ignoring the very complexity which ought to be at the heart of investigations into the spatiality of publicness (and privacy). Publics and privates, Mimi Sheller and John Urry (2003: 108) suggest, 'are each constantly shifting and being performed in rapid flashes within less anchored spaces'. As such, they believe that the complex 'where' (and 'when') of publicity and privacy cannot be captured by static 'regional' or topographical conceptions of public and private as distinct 'spheres' or 'spaces'. Indeed, they go so far as to argue that such conceptions of public and private ought to be consigned to the dustbin of history:

> Despite the heroic efforts of 20th century normative theorists to rescue the divide, the various distinctions between public and private domains cannot survive . . . [T]he hybridization of public and private is even more extensive than previously thought, and is occurring in more complex and fluid ways than any regional model of separate spheres can capture. Any hope for public citizenship and democracy, then, will depend on the capacity to navigate these new material, mobile worlds that are neither public nor private (Sheller and Urry 2003: 113).

Of course, not all conceptions of 'public space' are of the static and topographical variety to which Sheller and Urry object. There is also a tradition of defining 'public space' from a *procedural* rather than topographical perspective. Defined procedurally, 'public space' is understood to be any space which, through political action and public address at a particular time, becomes 'the site of power, of common action coordinated through speech and persuasion' (Benhabib 1992: 78). Drawing on Hannah Arendt (1958), Seyla Benhabib (1992: 78) argues that public spaces (defined in this procedural sense) can exist across 'diverse topographical locations'. Arendt was particularly influenced by the Greek conception of publicness, and while she recognized that in this conception publicness was strongly associated with the *polis*, she was quite careful to spell out the distinction between the *polis* and the physical spaces of the city:

> The *polis*, properly speaking, is not the city-state in its physical location; it is the organization of the people as it arises out of acting and speaking together, and its time-space lies between people living together for this purpose, no matter where they happen to be. 'Wherever you go, you will be a *polis*': these famous words became not merely the watchwords of Greek colonization, they expressed the conviction that action and speech create a space between the participants which can finds its proper location almost any time and any-where. It is the space of appearance in the widest sense of the word, namely, the space where I appear to others as others appear to me, where men exist not merely like other living or inanimate things but make their appearance explicitly (Arendt 1958: 198–9).

To the Greeks, in other words, 'not Athens, but Athenians, were the *polis*' (Arendt 1958: 195). Here, Arendt's concept of 'appearance' is not reducible to the 'visibility' associated with physical co-presence in a place which is so important in topographical approaches to 'public space'.

What are the implications of these procedural conceptions of public space for our consideration of the *urban* dimensions of public-making and public address? If publics can indeed find their 'proper location almost any time and anywhere', as Arendt would have it, then the 'public spaces' of the city (topographically defined) would appear to have no privileged relationship to public-making and public address. Rather than focusing only on these sites or places, procedural conceptions of 'public space' draw explicit attention to the complex geographies of publicness that topographical approaches struggle to capture. So, in its procedural conception, 'public space' is not reduced to a fixed set of topographically defined sites in the city which act as a kind of 'stage' for representation before a gathered 'audience'. Different forms of public address may take place in a park, and over a kitchen table, and as such both could be considered as 'public space' from a procedural perspective. And of course, 'public space' does not only take the form of such 'physical' sites. From the procedural perspective, we may consider various kinds of *media* as 'public space', given their central role in facilitating the formation of modern and contemporary publics. This approach is advanced by Clive Barnett (2004: 190), who is critical of 'geographers' determination to translate the public sphere into bounded public urban spaces of co-present social interaction'. For him, a process-based approach to the public sphere requires us to 'stretch out' our conception of the public to take into account the importance of a range of spatial practices to the making of publics, directing attention to the role of media and communications practices in particular. The historical emergence of print media in the form of regularly published journals and newspapers is often regarded as fundamental to the formation of publics in the modern period (see in particular Anderson 1983; Habermas 1989; Warner 1990). And of course, the twentieth century has witnessed the widespread diffusion of electronic media in the form of radio, television and more recently the internet. Sheller and Urry argue that new communications technologies have given rise to an important new space of publicness – Arendt's 'space of appearances' may be a *screen* on which 'private' lives are made 'public' and 'public' issues are transmitted into 'private' contexts:

> Where once 'staging' was the operative metaphor for public events, now 'screening' is more appropriate to describe those contexts where privacy has been eroded and where supposedly private lives are ubiquitously screened (Sheller and Urry 2003: 118).

Of course, if we accept that the gathering places of the city have no privileged relationship to public-making, it does not necessarily follow that these places have been rendered *irrelevant* as a form of 'public space', as some champions of new media and communications technologies are prone to argue. The literature on cyberspace, for instance, is littered with techno-utopian claims that it has replaced/is replacing the gathering-spaces of the city as the pre-eminent public space (for a critical discussion of these claims, see Robins 1996). Ironically, perhaps, such claims frequently deploy urban metaphors to make sense of the spatiality of cyberspace and the media (Crang 2000) – Paul Virilio's claim that 'The screen has become the new village square' is one case in point (quoted in Featherstone 1998).

Unfortunately, however, some of those who have leapt to the defence of the city in the face of techno-utopian claims about its irrelevance have done so by simple reassertion of the primacy of urban gathering-spaces over other forms of public space. In these counter-claims, the urban (or the 'real', the 'material') is often defined in *opposition to* the media (or the 'mediated', the 'virtual'), so that these two forms of public space appear locked in a battle for ascendancy whose outcomes will determine the very possibilities for democratic citizenship. But we should not defend the city's importance for public-making by reasserting the value of its gathering-spaces *over and against* other forms of public space. Mitchell's (2003) claim that the internet cannot match the street as a space of democratic representation is illustrative of the problems of such an approach. Quite rightly, Mitchell is wary of those who herald the internet as *the* new public space. In noting his concerns about the limitations of the internet as a public space, he argues that 'the material structure of the medium closes off political possibilities and opportunities' (Mitchell 2003: 145). This concern with the consequences of the 'material structure' of the internet is important, but it should be applied to *any* kind of public space, not least the streets which he champions as the privileged terrain of publicness. Surely *all* kinds of public space have a 'material structure' which influences the political possibilities and opportunities they afford? This is precisely the point to which our attention is drawn by procedural conceptions of public space – if publics have *no* proper location, then we should be wary of claims that *any* kind of space has a privileged relationship to publicness, whether they be 'squares' or 'screens'.

The challenge posed by procedural conceptions of public space, then, is to find a new way of conceptualizing the urban dimensions of public-making which avoids the tendency to either privilege or denigrate the city's gathering-spaces as 'public space'. To help us address this challenge, we can draw two important insights from the analysis presented so far in this section. First, if we accept that all forms of public space have a distinct 'material structure', then we ought to explore the particular materiality of different forms of space, asking about how this materiality is made and remade, and considering the consequences of this materiality for different forms of public address and for different publics.[6] In other words, while we might accept the notion that no space should be privileged as a form of 'public space' because publics have no proper location, this does *not* mean that all kinds of space are equivalent or equally available for those engaged in struggles to make publics. Rather, different kinds of public space offer different possibilities and opportunities for public action, and these differences require empirical analysis. Yes, an activist might turn a street, a kitchen or a website into a 'space of appearances' – but they are unlikely to do so without some appreciation of the different opportunities that these different spaces afford for public action. Nor are they likely to restrict their action to any one of these spaces. This leads me to my second point – we should not frame different kinds of public spaces as stark alternatives to one another. Rather, we ought to explore the ways in which publics *combine* a variety of 'public spaces' in their action. Indeed, we can use the example of protests which disrupted the 1999 meeting of the World Trade Organization in Seattle to illustrate this point. For Mitchell, these protests illustrate the ongoing primacy of the city's gathering-spaces for politics – the protests, he asserts, would have been nothing without 'people in the streets' (2003: 147). But what purpose is served by asserting the primacy of the street over the media (or vice versa) in such a case? Certainly, this is a distinction which many of the activists involved in the Seattle demonstrations did not make – these events were as significant for the development of new forms of activist media and tactical interventions in the mainstream mass media as they were for the occupation of the streets. The streets and the screens here are distinct spaces for public action, but actions undertaken through these distinct spaces took shape in close relation to one another. While many kinds of 'public space' exist, none exists in isolation – rather, these spaces develop and mutate in complex relation to each other. 'New' forms do not replace the 'old', but draw them into new combinations.[7]

Public address and 'public space': an initial summation

These observations about the distinct and yet relational materiality of different kinds of 'public space' (procedurally defined) raise some intriguing dilemmas for our inquiry into the urban dimensions of struggles over the making of publics. Drawing on the insights of procedural approaches to public space, our inquiry should not privilege those spaces defined as 'public' by topographical conceptions of the public/private distinction. However, we cannot simply reject or ignore these conceptions, as suggested by Sheller and Urry. Rather, we are now in a position to re-contextualize topographical conceptions of public and private within a wider inquiry into the materiality of different kinds of 'public space'. Certainly, we have established that being public is not simply a matter of being *in* public, where being in public is equated with establishing an embodied presence in a particular site in the city defined as 'public' in a topographical sense. And yet, the materiality of some procedurally defined 'public spaces' is nonetheless fundamentally shaped by norms which continue to invoke spatialized distinctions between 'public' and 'private'. In some senses at least, the public/private distinction *is* materialized in a topographical register, such that moving from 'public' to 'private' can be 'experienced as crossing a barrier or making a transition' because of the different kinds of visibility afforded by different kinds of place (Warner 2002: 26). To be flippant, while Sheller and Urry may argue against regional conceptions of public and private in the pages of an academic journal, they may not be inclined to extend their critique of regional conceptions of 'public' and 'private' so far as to advocate (or practise) masturbation in a 'public' street. Greek philosopher Diogenes is said to have done this repeatedly in ancient Athens, in a kind of 'performance criticism' of normative ideas about public and private (Warner 2002: 21). Perhaps Sheller and Urry would also be so bold, but surely neither they nor their audience would consider such an event unremarkable, precisely because it would transgress currently acceptable norms about 'public' and 'private' which have a strong topographical dimension.

So, it would appear that we have come full circle. The urban dimensions of struggles over the conditions of public address are not adequately conceptualized as struggles over access to public space, where 'public space' is understood exclusively in a topographical *or* a procedural sense. Both topographical and procedural approaches to public address and its relationship to 'public space' point us towards the geographical complexity of publicness in its different forms – as a context for action, a kind of action, and a kind of actor. However, topographical approaches mistakenly see a direct equivalence between these three dimensions of publicness. While procedural approaches capture some of the dynamic geographies of public address, they fail to appreciate fully the persistent power of normative topographical mappings of public and private. Each approach, in other words, captures a particular aspect of the relationship between publicness and the city at the cost of neglecting other important aspects – the relationship is more complex than either seems to allow. I agree with Weintraub (1997: 3) that this 'complexity needs to be acknowledged, and the roots of this complexity need to be elucidated'. As the discussion above illustrates, when different fields of discourse about publicness are allowed to operate in mutual isolation, or when their categories are casually or unreflectively blended, confusion or even absurdity can be the result (Weintraub 1997: 2–3). To find a way through these murky waters, we need to have a much clearer appreciation of the multidimensional nature of the public/private distinction and its various applications across different realms of social life.

Public Address and the City

Analyses of the public/private distinction provided by Benn and Gaus (1983) and Weintraub (1997) are particularly useful in dissecting the complexities revealed in the discussion of different conceptions of 'public space'. As Weintraub (1997: 1–2) has observed, use of the conceptual vocabulary of 'public' and 'private' in reference to urban space (and to other domains of social life):

> often generates as much confusion as illumination, not least because different sets of people who employ these concepts mean very different things by them – and sometimes, without quite realizing it, mean several different things at once.

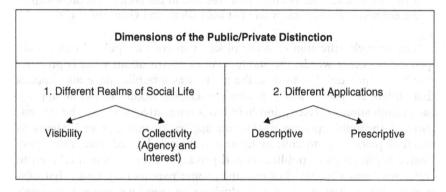

Figure 1.1   Dimensions of the public/private distinction

There are two sets of meanings attached to the public/private distinction that are particularly important for our discussion of the urban dimensions of public-making (see Figure 1.1). First, the public/private distinction is used in reference to *distinct realms of social life*. When we describe actions taking place 'in public' or 'in private', public and private are understood as different contexts for action with different forms of *visibility*. In this sense, the conceptual vocabulary of public and private is used to distinguish between what is open, revealed or accessible (i.e. public), as opposed to what is hidden or withdrawn (i.e. private) (Weintraub 1997: 5). When we describe actions which are taken by 'a/the public' and/or actions taken in the 'public interest', publicness is understood with reference to the *collectivity* of different forms of *agency* and *interest*. In this sense, the conceptual vocabulary of public and private is used to distinguish between what is collective, or affects the interests of a collectivity of individuals, versus what is individual, or pertains only to an individual (Weintraub 1997: 5; Benn and Gaus 1983).[8]

Second, uses of the public/private distinction are further complicated because there are at least two ways in which the public/private distinction is *applied*. It is variously used as a *descriptive* device and as a *prescriptive* device. Both of these applications of the public/private distinction are fundamentally normative (Benn and Gaus 1983: 11–12). It might seem relatively obvious that prescriptions involving publicness and privacy have a normative content. For example, ordinances against nudity in public space clearly invoke norms about what constitutes appropriate behaviour in places where one's body is visible to others. But descriptive applications of publicness or privacy are no less normative:

> to describe an object as private [or public] implies that it satisfies some, at least, of a bounded set of conditions specified in the norms, without which the normative implications would not hold (Benn and Gaus 1983: 12).

The very classification of some place or interest as 'public' rather than 'private', in other words, inevitably invokes norms about what is properly 'public' or 'private'. So, to describe a street as a public space also implies that streets are places where norms proscribing nudity in public apply – the classification or description here has a normative content. The normativity of the public/private distinction applies to interests and agents as much as places (i.e. to collectivity and visibility). Indeed, sometimes normative applications of publicness and privacy combine criteria relating to collectivity and visibility. For example, some parents may claim that how they discipline their children in the 'privacy of their own home' is no one's business but their own. Here, the ideological classification of the home as a private space is used to protect it from institutional sanctions against physical disciplining of children, which are seen to erode that inherent privacy and its associated rights. The 'privacy' of the home has been publicly challenged by advocates of children's welfare and rights, who argue that the nature of interactions between parent and child within the home is a matter of 'public interest'. Because distinctions between public and private are essentially normative in nature, they have been a matter of political and theoretical contention. In particular, the distinctions articulated in the liberal political tradition which render some matters/bodies/spaces/actions/etc. as private and others as public, and which have been widely institutionalized in politics and law, have been the target of concerted political action (Warner 2002: 39).

We are now in a position to offer a more refined diagnosis of the problems with the topographical and procedural approaches to 'public space' which have dominated thinking about the relationship between publics and the city. Put simply, topographical models of public space use 'public' to denote spaces of sociability in the city where one's actions are visible to others, while procedural models of public space use 'public' to denote spaces where one may take part in collective discussions about common interests and issues. Each approach draws attention to an important dimension of publicness. But each also fails to trace fully the complex interactions between the distinct dimensions of publicness, either simplifying or neglecting the nature of these interactions.

Topographical conceptions of public space usefully draw our attention to the power of regional distinctions between public and private which persist in the form of socio-spatial norms about conduct and action in (certain parts of) the city. That is, regimes of place often invoke norms about what behaviour is appropriate 'in public' and 'in private' in order to foster particular forms of conduct. Of course, these norms are contested and change over time and space. As such, struggles over the forms of conduct which are normalized in particular urban time-spaces often take the form of struggles over the terms of accessibility of 'public space'. Topographical conceptions of 'public space' equate being *in* public with being public in its collective sense. Publics (as collectives) and public action are not contained within spaces typically mapped as 'public' in a topographical sense. Procedural conceptions of 'public space' draw attention to the dynamic geographies of publicness as collective interests and agency, which do not conform to the conventional mappings of public and private. Nonetheless, we cannot simply choose to do away with topographical or regional conceptions of 'public' and 'private'. It may be true that one can address a public from one's bedroom as well as from a street corner, and indeed that one's 'private life' may be publicly 'screened'. But this does not mean that conventional designations of the bedroom as 'private' and the street as 'public' no longer have any power at all. The distinction between public and private cannot therefore be reduced wholly to a procedural distinction. As Michael Warner (2002: 28–9) has argued:

> attempts to frame public and private as sharp distinction or antimony have invariably come to grief, while attempts to collapse or do without them have proven equally unsatisfying.

The challenge, then, is to build a framework for investigating the urban dimensions of public-making which is sensitive to the multidimensionality of publicness and privacy.

Building a framework that is sensitive to the multiple dimensions of publicness and privacy will also help us to bring together the different dimensions of 'the city' that are privileged by topographical and procedural approaches to public space. In topographical approaches to public address and public space, 'the city' features as a network of physical sites which serve as a stage for public representation and visibility. In procedural approaches to public address and public space, 'the city' features more as a kind of 'being together' that is as much a matter of public deliberation and collective concern as physical propinquity. Both of these approaches bring out distinct but related urban dimensions of public-making. We can see both of these dimensions of 'the city' at work in Upski's efforts to 'find an audience'. While Upski might have found some people on a street corner, or by posting a newsletter to a pillar on a train station platform, he found others through book sales in record shops and online retailers. I became part of Upski's public when I found his book in an independent bookstore while on a trip to the United States some years ago. 'The city' still played a vital role in making this connection possible – but not in the form of a physical 'public space' where I could witness one of Upski's talks. Upski and I connected through a *shared interest* in the state of contemporary cities, rather than through *sharing a space in* one of those cities (how could a book called *Bomb the Suburbs* not leap off the shelves for an urban researcher with an interest in graffiti?).

The Structure of this Book

In the next chapter, I take up the challenge of developing a framework for research into the urban dimensions of public-making which is sensitive to the multidimensionality and complexity of publicness and the city. This framework is developed by establishing a conversation between urban studies and critical social theories of the political 'public sphere'. The distinct trajectories of these two literatures has meant that studies of publicness in the 'polis' and in 'print' have mostly failed to connect (Iveson 2003; Smith and Low 2006). But the connections are there to be made. If the public space debates in urban studies have tended to lack conceptual clarity with regard to what constitutes 'publicness', then it is also true that the spatial vocabulary of critical social theory remains underdeveloped in some important respects.

The framework developed in Chapter 2 is then applied to investigate a series of struggles over the urban dimensions of publicness over the next five chapters. The conceptual organization of these case studies is discussed further in Chapter 2, but let me now offer some preliminary orientation concerning these chapters. They each explore the ways in which the urban is used and produced in struggles to establish particular forms of publicness in different Australian cities. The chapters range from considerations of political protesters seeking to use the grounds around Parliament House in Canberra to young people hanging out on the streets of inner-city Perth and writing graffiti in Sydney, from a coalition of women and their supporters mobilizing to keep men out of a public swimming pool in Sydney to men cruising the parks and public toilets of Melbourne for sex with other men. These studies are all based on fieldwork conducted at various points over the past eight years.

In seeking to convince readers in Australia and beyond that these studies might be of interest, I make no claim that they are *representative* of struggles over the making of publics which take place in other cities in Australia or indeed in other parts of the world. The struggles I consider are neither 'typical' cases nor are they fought over 'paradigmatic places' whose present might become someone else's future. Nonetheless, as the title of this book suggests, I do hope that these investigations into particular publics in particular cities might also be *illustrative* of the relationship between publics and the city more generally. Michael Warner (2002: 11) has argued that 'the idea of a public has a metacultural dimension; it gives form to a tension between general and particular that makes it difficult to analyze from either perspective alone'. The same, I think, could be said about the idea of a city. I share Jennifer Robinson's (2002: 549) view that urban theory might also benefit from giving a little more consideration to 'the difference the diversity of cities makes to theory'.

In the final chapter, I return to consider the conceptual and political implications of the approach to publics and the city that I develop over the course of the book. My main claim here is that a revised concept of publicness can still be a powerful tool for critical analysis of contemporary urbanization.

Chapter Two

Publics and the City

The image used on the cover of this book is of course a bit of a joke. I won't be promoting *Publics and the City* with pole-posters announcing its publication. Nor do I expect to be reading excerpts at train stations or on street corners, Upski style. This is because I don't really imagine reaching my intended audience in this way. I hope that this book might contribute to ongoing discussions about the nature of urban life and modern publics that are conducted in universities by academics and students. These discussions are happening across a range of academic disciplines which have an interest in such matters, such as geography, urban planning, sociology and anthropology. I've already made a calculation that a book published by an academic publisher might be an appropriate form of address to contribute to these discussions. The publishers have asked me to provide a list of other academics who I think might be interested in the book, to whom they can send notice of the book's publication. But hopefully, the book's circulation will not be restricted to people I already know. The book might reach others by making an appearance on Blackwell's table at an academic conference, or in one of the booklists they provide to academics and university librarians. If anyone reads the book and thinks it has something useful to say (or indeed if they think I have got it all wrong!), they may cite the book in a paper published in an academic journal. A reader of that journal paper may care to follow up the reference to this book. Of course, all of this is speculation on my part – speculation informed by experience of academia, but speculation nonetheless.

The fact that the image of a pole-poster promoting this book seems incongruous points to an obvious but nonetheless significant feature of public address. Public address can take a variety of different forms. These forms have both material and imaginative dimensions – in addressing a public, we produce and we mobilize different genres, styles, technologies, rhythms and spaces, and this requires us to imagine the prior existence of an audience who might understand (and be receptive to) the form and content of our address. Publics are thus *social imaginaries* (Taylor 2004; Warner 2002).

In this chapter, my purpose is to develop a framework for inquiry into the diverse forms of public address and the relationship between the social imaginaries of publics and spatial imaginaries of 'the city'. The chapter proceeds in three sections. First, I expand upon my understanding of *public address* and *public spheres*. Here, I sketch out the nature of public spheres as social imaginaries, and then proceed to outline a framework for thinking about the relations between publics and 'counter-publics' that draws on recent interventions in public sphere theory derived primarily from feminist and queer theorists. Second, I ask how this understanding of public spheres might relate to the city. Here, my purpose is to outline a framework for inquiry into struggles over the production of public urban geographies. This framework is intended to rescue the city from its almost exclusive association with gathering-spaces and propinquity in considerations of the urban dimensions of public address (see Chapter 1, and also Calhoun 1998). I argue that there are three distinct but related ways in which cities play a part in the formation and interaction of public social imaginaries – as venues of public address, as objects of public debate and connection, and as collective subjects which serve as the common horizon for diverse publics. Finally, I offer brief descriptions of the case studies to follow, positioning them within the complex and dynamic relationship between these aspects of urban publicness.

Public Spheres and Public Address

When the Australian prime minister seeks to address 'the Australian public', how does he do it? He does not talk to every Australian directly and personally. Rather, he circulates some kind of 'text' (be that a policy document, a press briefing, a speech to Parliament or a business association) through channels which he hopes might enable him to access 'the Australian public'. In this first part of the chapter, I want to offer some preliminary reflections on the nature of public address – that is, address which is oriented towards a 'public sphere'. I use the term 'public sphere' to refer to an arena of self-organized discursive interaction in which common interests, values, experiences and desires are posited and contested by people who do not address each other personally. My focus here is on the *production* of public spheres through forms of address which take the very existence of common interests and shared worlds as both their premise and their problem.

'Public address' and 'public spheres' are fundamentally dependent on each other. This relationship has a 'kind of chicken-and-egg circularity' with profound consequences (Warner 2002: 67). Michael Warner's (2002) discussion of public spheres is particularly useful in its identification and clarification of this circularity. As he argues, while public address takes for granted the existence of a public sphere which can be addressed, these public spheres cannot exist without instances of public address. These instances of public address are oriented towards a horizon of strangers which must be imagined, because these strangers are not addressed personally or directly.[1] Rather, they are addressed as fellow participants in a scene which facilitates the circulation of texts among strangers through 'venues of indefinite address' (Warner 2002: 86).[2] Unless one can imagine the existence of channels of communication which act as venues of indefinite address (as John Howard does when he addresses a gathering of journalists or parliamentarians), one's address must remain personal and direct rather than public. And unless one's text actually reaches people who understand themselves as one of the strangers on the horizon which is being addressed, that public sphere fails to materialize. As such, public spheres are social imaginaries that are always in the making.

Because they are always in the making, the 'shapes' of public spheres have varied significantly across different historical and geographical contexts (Carpignano 1999). Participants in public spheres may mobilize a range of forms of public address in efforts to find their audience. As Warner (2002: 7) has noted, public address takes an almost infinite variety of forms:

> Texts cross one's path in their endless search for a public of one kind or another: the morning paper, the radio, the television, movies, billboards, books, official postings. Beyond these obvious forms of address lie others, like fashion trends or brand names, that do not begin 'Dear Reader' but are intrinsically oriented to publics nonetheless ... Your attention is everywhere solicited by artefacts that say, before they say anything else, Hello, public!

Indeed, the forms of public address which seek our attention are in a constant state of mutation and multiplication. But precisely because instances of public address must have a specific form, every public sphere has its limits. The venues of indefinite address that enable the production of public spheres are not infinitely accessible. Even if one does not know exactly who will be able to access a given public address, any venue of indefinite address still 'selects participants by criteria of shared social space' (Warner 2002: 106). To access John Howard's address in the form of a newspaper, for instance, requires me to be in the habit of reading newspapers, to have a grasp of the language of the newspaper in question, to have the money to buy the newspaper (or the time to go the library, or the luck to find one left on the seat of my train home, etc.). Such criteria of the shared social space of a newspaper (and other forms of public address) are a product of the circular process that produces public spheres – in order for them to circulate, texts must take particular forms, but such forms also limit the scope of their circulation. As Warner (2002: 106) puts it:

> These criteria inevitably have positive content. They enable confidence that the discourse will circulate along a real path, but they limit the extension of that path.

Not surprisingly, the criteria which enable and limit access to scenes of public address have been of great interest to critical theorists. Their research has often focused attention on the nature of these criteria in particular empirical contexts, and their theoretical reflections have often sought to establish grounds for normative judgement about the limits they help to establish. In their own distinct ways, for example, Arendt's (1958) and Habermas's (1989) now classic studies sought to develop a set of normative principles for democratic public spheres by critically interrogating the limits and unrealized possibilities of historical public spheres in ancient and modern European contexts. More generally, critical theorists have argued that the 'shape' of a public sphere might work to place limits on the kinds of people who can access its shared social spaces, the styles and forms that public address can take, and the topics that are considered proper matters for public discussion. As such, these shapes have vital political consequences – the limits of a given public sphere establish paths or channels of discourse that organize 'partitions of the perceptible' (Rancière 1999: 25) which deny some people or issues access to its shared social space. When such limits are made to appear 'normal' through institutional arrangements premised on the notion that they are natural or pre-political, this is of particular concern. For example, feminists have identified and critiqued contexts in which women have been excluded from venues of public address such as parliamentary politics, where women's ways of speaking have been deemed to be too 'emotional' or 'irrational' to be admitted as a legitimate form of public address, or where issues of concern to women such as violence in the home have been deemed to be matters of 'private' rather than 'public' interest. Indeed, Carole Pateman (1983: 281) observed some years ago that:

> The dichotomy between the public and the private is central to almost two centuries of feminist writing and political struggle; it is, ultimately, what the feminist movement is about.

It is of course possible that blockages preventing access to a given public sphere are not a product of the *principles* underpinning a public sphere, but rather a product of their incomplete *application*. If this were the case, then the job of critique is to hold a mirror up to actually-existing public spheres, arguing for expansion on the grounds that a public sphere live up to its own norms of participation. Debates over the limits of bourgeois public spheres in early modern Europe provoked by Habermas's (1989) account of their development and structural transformation are a case in point. These public spheres, which first emerged in the world of letters, were mobilized by the bourgeoisie in their struggle against absolutist rule. They proposed that 'public opinion' – the product of private people coming together and deploying their 'reason' through rational-critical debate – could also be used as a means to legitimate the actions of the state. This crucial role for public opinion was justified on the grounds that the public sphere was open to all, regardless of status based on grounds such as birthright or patronage:

> The bourgeois public's critical debate took place in principle without regard to all pre-existing social and political rank and in accord with universal rules (Habermas 1989: 54).

Habermas was well aware that workers without property and women (among others) were denied status as participants in bourgeois public spheres. As he noted, 'the positive meaning of "private" emerged precisely in reference to the concept of free power of control over property that functioned in a capitalist fashion' and it 'had its home, literally, in the sphere of the patriarchal conjugal family' (Habermas 1989: 74, 43). Such restrictive qualifications for access were supported by political-economic theories which purported to show that in a free market economy, anyone could acquire property and education if they so desired. Nonetheless, Habermas identified a positive potential in the principles of bourgeois public spheres which he believed to be unrealized. While the European bourgeois public spheres he studied clearly privileged particular interests, they simultaneously opened up the possibility of the formation of a 'universal interest everyone can acknowledge' through rational-critical debate in which contributions were assessed on the basis of reason rather than status (Habermas 1989: 235). In the end, however, Habermas concludes that this potential was squandered. Liberal responses to the limits of the bourgeois public sphere successfully sought to widen the accessibility of the public sphere to larger numbers of people while reducing the role of the public sphere to selecting political representatives rather than fully deliberating on matters of state: 'the *principle* of the public sphere, that is, critical publicity, seemed to lose its strength in the measure that it expanded as a *sphere*' (Habermas 1989: 140, original emphasis).

Habermas's critics have argued, however, that the limits of European bourgeois public spheres were not simply a matter of the incomplete application of the principle of being 'open to all' regardless of status. The extension of the principles of the bourgeois public sphere to workers, women and others was still premised on a fundamental subordination – that participants would be able to put aside their 'private' differences in status in order to debate the public or common interest. This still blocks the consideration of questions of subject formation and identity from the scene of public address and debate. These terms of inclusion had the effect of:

> excluding some of the most important concerns of many members of any polity – both those whose existing identities are suppressed or devalued and those whose exploration of possible identities is truncated (Calhoun 1997: 83).

Thus, as Nancy Fraser (1992: 115) put it:

> There is a remarkable irony here, one that Habermas's account of the rise of the public sphere fails to fully appreciate. A discourse of publicity touting accessibility, rationality, and suspension of status hierarchies is itself deployed as a strategy of distinction.

From this perspective, the terms on which a public is 'open' are crucial – as Pateman (1983: 68) concludes, 'access is not enough'.

If 'access is not enough', then what? Those who are marginalized by the shape of a sphere of public address may try to overcome such limits through the creation of new discursive scenes which embody different possibilities for public address. Such scenes can work to open up a new 'partitions of the perceptible'. They expand the range of people who have access to public spheres, and expand the available styles and topical concerns of public address. When such publics have embodied an explicitly oppositional stance in relation to prevailing notions of what counts for 'normal' forms of public address, they have come to be known as *counterpublics*. Nancy Fraser's definition of counterpublics has been particularly influential:

> members of subordinated groups – women, workers, peoples of color, and gays and lesbians – have repeatedly found it advantageous to constitute alternative publics. I propose to call these *subaltern counterpublics* in order to signal that they are parallel discursive arenas where members of subordinated social groups invent and circulate counterdiscourses to formulate oppositional interpretations of their identities, interests, and needs (Fraser 1992: 123, original emphasis).[3]

If this is the case, then the priorities for critical theory must shift – rather than holding on to the principles of existing public spheres as the basis for challenging their limits, we ought to be more concerned with the *production* of counterpublic spheres which enable novel scenes of public address to develop (Negt and Kluge 1993: 79).

The production of counterpublic spheres involves the same processes as those described above for the production of publics more generally. The same circularity applies to public address in counterpublic spheres. In other words, the only way through blockages to different forms of public address in existing public spheres is through the construction of a shared social space not characterized by those blockages. And yet the construction of such a new scene itself relies on instances of public address – counterpublic spheres are thus both a *product* and a *prerequisite* of new forms of discursive inter-action (Negt and Kluge 1993: 94). So, for example, the development and public circulation of critiques of women's exclusions from some public spheres has itself been achieved through the building of feminist counter-public spheres with their own forms of public address. These instances of public address are only possible because of the possibility of imagining a horizon of strangers to whom they might circulate more broadly. And so the cycle goes, allowing the emergence of a new 'horizon of experience' which relates to a particular 'context of living' (Negt and Kluge 1993).

The 'counterpublic horizon', like the horizon of other publics, is not infi-nite even if it is indefinite (Warner 2002). However, the 'distance' of this horizon is a constant problem for participants in counterpublic spheres, because participation in a counterpublic sphere is likely to be risky. A range of institutional interventions tend to frame different ways of being public hierarchically – so that some are assumed to be 'normal' while others are marked as 'deviant' or 'disorderly'. That is to say, 'ordinary people are presumed not to want to be mistaken for the kind of person who would participate in this kind of talk or be present in this kind of scene' (Warner 2002: 120).[4] The strangers on the horizon of instances of counterpublic address might be indifferent, sympathetic or hostile to its style and content, and to the people involved in its circulation. As such, when one addresses a counterpublic, one addresses the *particular* strangers who are fellow par-ticipants in that public while simultaneously knowing that one's address might also reach other strangers (Warner 2002: 120). The production of counterpublic spheres inevitably embodies this tension in some form or another – indeed, the nature of the relation between counterpublic spheres and other publics will vary across time and space with different effects. Some participants in counterpublics may seek to maintain an 'open horizon', in the hope of expanding the scope of their scene and dissolving the existing limits of other scenes, and yet this approach risks hostile interventions which seek to close down opportunities for counterpublic interaction. Other participants may seek to police the boundaries of their public sphere, establishing a counterpublic as a 'separate camp' with no relation at all to other publics. Yet this approach sets up exactly the kinds of problem inherent to any public which sets limits to participation based on universal and abstract value judgements rather than open interaction (Negt and Kluge 1993: 62–3).

Contests over the coexistence of publics and counterpublics are often conducted with reference to some wider concept of '*the* public' – a concept which serves as an 'imaginary convergence point that is the backdrop of critical discourse' among different publics (Warner 2002: 55). That is to say, participants in some scenes of public address may claim that their deliberations are the best or only means to secure 'the public interest', while framing the actions and interests of others as 'particular' or 'parochial'. As Negt and Kluge have observed (1993: xlviii), 'this general over-riding public sphere runs parallel to these fields as an idea, and is exploited by the interests contained within each sphere'. Indeed, it is all the more likely to be mobilized by those publics where participants can take their particular forms of public address for granted, 'misrecognizing the indefinite scope of their expansive address as universality or normalcy' (Warner 2002: 122). 'The public', here, is 'the people' as a social totality. In the face of the inevitable and constant failure of any public to embrace the social totality, however, any claim that a scene of public address can represent 'the public' is always open to challenge. Such challenges are of course a problem for any institution, group or individual which claims to legitimate their actions on behalf of 'the public'.

Drawing on the experiences of different counterpublics, some critical theorists have sought to develop normative theorizations of how 'the public' would look if different publics were not framed against one another hierarchically. How might publics interact in such a 'heterogeneous public sphere'? Establishing new forms of non-hierarchical interaction would first require a mutual recognition across different publics that *the* public sphere in fact constitutes a 'structured setting where cultural and ideological contest among a variety of publics takes place' (Eley 1992: 306). The construction of such a wider horizon of interaction across difference, it is argued, remains crucial if the shared problems of living together in a given polity are to be politically addressed:

> Recognizing the existence of multiple public spheres thus is not an alternative to asking many of the questions Habermas asks of *the* public sphere, i.e., about public discourse at the largest of social scales and its capacity to influence politics. It simply suggests that these questions need to be answered in a world of multiple and different publics (Calhoun 1996: 460: original emphasis).

To the extent that the wider horizon of discourse across multiple publics might be oriented towards a 'common good' in a given polity, this common good:

> *can* be interpreted simply as the addressing of problems that people face together, without any assumption that these people have common interests or common way of life, or that they must subordinate or transcend the particular interests and values that differentiate them (Young 2000: 40).

Iris Marion Young (1990: 120) has offered two political principles as basic ground rules for how such deliberation ought to take place:

> (a) no persons, actions, or aspects of a person's life should be forced into privacy; and (b) no social institutions or practices should be excluded *a priori* from being a proper subject for public discussion and expression.

In offering such principles, Young and others (see for example Benhabib 2002) are asserting that *the* public must itself be understood as both a prerequisite *and a product* of public address, in the same manner as the different publics from which it is constituted. Before engaging with these current debates any further, I now want to address directly the spatiality of public social imaginaries, in particular the urban geographies of public address.

Public Address and Public Urban Geographies

Vital to the production of public spheres is a certain temporality, a rhythm which sustains the social imaginaries that support public address. The regular (usually weekly or daily) publication of newspapers in the early modern period, for example, allowed readers to imagine themselves participating with others whom they did not know in a shared, ongoing dialogue about public affairs (Anderson 1983; Taylor 2004; Warner 2002). What of the spatiality of public spheres?

It will have become apparent in the previous section of this chapter that the existing literature on public address and public spheres draws heavily on a range of spatial concepts and metaphors – the most obvious of which is of course the concept of a 'sphere', described above variously as 'horizons of experience' (Negt and Kluge 1993) or 'social spaces' (Taylor 2004; Warner 2002) of discursive interaction with distinct 'shapes' (Carpignano 1999). Most commentators on public address have made it very clear that public 'spheres' are not to be confused with territory or places as such – the limits of public spheres are not lines on a map. While common uses of 'public' and 'private' to refer to distinct spaces abound (see Chapter 1), the notion of a public sphere is distinct from such uses. Even if 'the public sphere' conjures up images of particular places (the parliaments, or the streets), things are not so simple. Anne Phillips' (1993: 93) ruminations on the spaces of politics could equally apply to spaces of publicness – 'perhaps the noun still attaches itself to definite places, while the adjective will go any old where?' Indeed, some have argued that topographical connotations of the term 'public sphere' are in large measure a consequence of the difficulties of translation. While the concept is often associated with Habermas's 1962 book *Strukturwandel der Öffentlichkeit* which was published in English in 1989 as *The Structural Transformation of the Public Sphere*, Warner (2002: 47) has argued that:

> The 'sphere' of the title is a misleading effect of English translation; the German *Öffentlichkeit* lacks the spatializing metaphor and suggests something more like 'openness' or 'publicness'.

Similarly, the translator of Negt and Kluge's 1972 book *Öffentlichkeit und Erfahrung* (published in full in English in 1993 as *Public Sphere and Experience: Towards an Analysis of the Bourgeois and Proletarian Public Sphere*) worried that while 'public sphere' seemed to capture their use of the term *Öffentlichkeit* as a 'spatial concept denoting the social sites or levels where meanings are manufactured, distributed and exchanged', it failed to capture adequately the other dimensions of their use of the term – to refer to both the ideational substance produced in these sites and the 'general horizon of social experience' to which discursive interaction is oriented. Initial translations of excerpts from the book sought to 'rehabilitate' the term *publicity* in place of *public sphere*, despite its other problematic connotations in English (see Peter Labanyi's note in Negt and Kluge 1988: 60).

Nonetheless, as Frederic Jameson (1988: 155) noted, while the notion of a public 'sphere' might be an imperfect translation into English of the German *Öffentlichkeit*, it 'generates interesting theoretical problems in its own right (which the term would not do in German)'. The public social imaginary has spatial as well as temporal dimensions. Just as the rhythms of publication and the temporalities of citational practices are essential to the production of public spheres, there is also a distinct spatiality at play – what Sennett (1978) called a *public geography* (although we shall have cause to disagree with the way in which Sennett understood this geography). The calculations which are involved in forms of public address, such as knowing 'whom to speak to and when and how, carry an implicit map of social space, of what kinds of people we can associate with in what ways and in what circumstances' (Taylor 2004: 25–6). Just as one reads or writes to the newspaper knowing it will be published tomorrow, so one knows that there are spaces (including mediatized spaces such as newspapers) through which one might access other members of 'the public'.

In some of most influential accounts of modern public spheres produced in the latter half of the twentieth century, *the city* figured as a kind of stage for particular forms of address and association. Most notably, cities appear in some accounts through the provision of places where people could imagine they were *in public* – most notably in places of collective discussion, in places where crowds could gather, and in places where people were exposed to others as strangers. In *The Fall of Public Man*, Richard Sennett (1978) sought to reveal the ways in which cities brought people into proximity with one another *as strangers*, thereby providing spaces which served as the stage for public life. Habermas's (1989) account of the bourgeois public sphere drew attention to the importance of domestic architecture in establishing distinct zones where private reason could be developed and deployed:

> The line between public and private sphere extended right through the home. The privatized individuals stepped out of the intimacy of their living rooms into the public sphere of the *salon*, but the one was strictly complementary to the other (1989: 45).

Famously, Habermas also regarded the coffee shops and tea-rooms of early modern cities to be crucial venues for private individuals to come together and make use of public reason. In their consideration of proletarian public spheres, Negt and Kluge (1993: 267) noted that in response to the emergence of collective concerns (such as concerns with government policies or workplace disputes), there is often a 'collective drive to the relevant places'. Sometimes the 'relevant place' might be a central city square, while in other instances it might be the factory gates.

Each of these thinkers bemoaned what they took to be the erosion of such possibilities in the contemporary cities at the time they wrote. Sennett (1978) identified and condemned what he saw to be the increasing importance attached to intimacy over estrangement in urban life, and has continued to worry about this trend (Sennett 1994). Habermas saw new principles for urban development, as well as other factors such as the emergence of a commercial mass media, as contributing to the structural transformation of the public sphere through a destruction of existing relationships between public and private spheres:

> The resulting configuration does not afford a spatially protected private sphere, nor does it create free space for public contacts and communication that could bring private people together to form a public (1989: 158).

In the streets and other gathering-places in particular, one might be 'in public' without engaging in public action:

> the individual satisfaction of needs might be achieved in a public fashion, namely, in the company of many others; but a public sphere itself did not emerge from such a situation (1989: 161).

Negt and Kluge condemned the hollowing out of post-war inner cities for eroding the possibilities for politics, just as Haussman's redevelopment of Paris had done a century before:

> the superimposition of administrative and banking centers with their sky-scrapers upon urban structures has brought about a similar destruction of the public sphere (Negt and Kluge 1993: 268n).

Up to a point, these accounts are suggestive of how we might approach the urban spatiality of public address. In particular, each focuses on the necessary *imaginative* dimension of being together with others in the city in a public fashion. That is, being public is not simply a matter of associating or gathering with others in a particular place. It is also a matter of imagining oneself and those others to be part of a public which exists beyond the spatial and temporal limits of any particular association or gathering. This inevitably provokes the need for more serious analysis of how being 'in public' (publicness as visibility) might relate to being part of a/the public (publicness as collectivity). However, as with the analysts of 'public space' considered in Chapter 1, when most accounts of the public sphere and public life turn to the city, they are overwhelmingly concerned with the city as a space of visibility (in particular, as the venue for simultaneous co-presence with strangers). Those other distantiated or de-centred (Young 2000: 46) 'venues' of public address which are considered necessary to sustain publics are, by contrast, typically associated with media (be it print, radio, television or the internet).

This association of the city with spaces of co-present sociability in accounts of the public sphere must give way to a more expansive understanding of the urban dimensions of public address and public-making. The metaphor of the city as a stage in fact limits our appreciation of the distinct (and yet related) urban dimensions of public social imaginaries. The multiple dimensions of publicness – as visibility and as collectivity (relating to both interest and agency) – can each take shape through 'the city', so long as we appreciate that the city is not (only) a physical space or container for action. Addressing a public *does* require a capacity to imagine the spaces through which one might appear to strange others – and this spatial imaginary is itself multidimensional. I now want to sketch out three distinct urban dimensions of this spatial imaginary. First, I show how publics may make use of (different places in) cities as *venues for public address*. Second, I demonstrate how cities and places in cities are also imagined and produced in struggles to make them common *objects of public debate*. Third, I consider the ways in which normative visions of 'the city' can also serve as a representation of 'the public', acting as a shared horizon for the interaction of distinct publics. As I hope to show, once we have identified and unpacked these dimensions, this opens up potentially fruitful enquiries into the tense and sometimes disjunctive relations between these dimensions of public-making.

1. Urban sites as venues of public address

A variety of urban sites can be mobilized as *venues* of different kinds of public address – that is, different kinds of places in cities can be put to work as 'venues of indefinite address' to establish 'scenes of circulation'. Public texts of one sort or another circulate through streets, parks, cinemas, bookshops and newsagents, homes, cafés and internet cafés, libraries, town halls and parliamentary buildings, convention centres, university seminar and lecture rooms, workplace noticeboards, public transportation facilities – the list could go on. Some of these venues conform to topographical conceptions of 'public space', and others do not. Each venue can potentially facilitate a variety of forms of public address and circulation. Stand on any busy street corner, for example, and you might find yourself addressed publicly in all manner of ways: by someone offering a speech, by a message on the t-shirt of a fellow pedestrian, by a billboard advertisement, by a public information notice or a sticker on a lamppost, by signs in shop windows, by sky-writing overhead, by a stencil on the footpath, by the radio you are listening to on your personal stereo, perhaps by a screen mounted on a nearby building, just to mention a few. In a bookshop you might find a good book, listen to a talk provided by an author, discuss the merits of a book with a sales assistant or fellow members of a book club, and read notices of forthcoming public events by looking at a noticeboard or picking up a flyer. Some forms of public address might be particular to their venues, while others can potentially make use of a range of venues.

The scenes of circulation established in and through urban sites and locations may involve some form of embodied co-presence. The potential of different urban sites to be mobilized as venues of public address is therefore in part determined by their accessibility to bodies and their hospitality towards various embodied activities. This is of course one of the reasons why the accessibility of various sites in the city has always been such a matter of contention. As noted in the previous chapter, some commentators are particularly concerned about the exclusionary regimes of access in sites like streets and parks that they perceive to be associated with emergent forms of 'neo-liberal' or 'advanced liberal' forms of urban governance (see for example MacLeod 2002; Mitchell 2003; Rose 2000).

Nonetheless, we should not make the mistake of thinking that urban public address is simply embodied public address which relies on simultaneous co-presence at a given site. This mistake is sometimes a feature of arguments which are mounted in defence of openly accessible urban public space.[5] The city's contribution to public address is not reducible to the gathering of crowds who are physically present to witness some kind of address. In fact, the circulation of different kinds of texts through sites in the city can involve a range of different configurations of embodied presence, and the actions of these bodies are often (if not always) mediated and mediatized in some fashion. Let me return to some different forms of public address on the street corner and in the bookshop to demonstrate some of these possible configurations and mediations.

First, let's consider a form of public address in the city which is almost paradigmatic – a verbal exchange between a speaker and an audience who are physically co-present in a particular place. A street-corner speech or a talk by an author at the bookshop are instances of embodied co-presence. Such embodied encounters between people co-present in a given site are still mediated – which is *not* to say determined in advance – by some-bodies' expectations of other-bodies. That is, we do not participate in encounters with others with a complete absence of knowledge. Rather, those others are in some ways 'known', even if they are known as 'unknown' or 'strange'. We carry expectations about others based on a range of factors, including prior encounters in similar situations and narratives about safety and danger in the city, among other things (Ahmed 2000). Think of a street-corner address from the perspective of the speaker and passers-by. The speaker might regard passers-by on a given street as a potentially sympathetic audience who would not otherwise access a message, or as a potentially hostile audience who could at least be jolted out of their comfort-zone by someone with an alternative perspective on the world getting in their face. Passers-by, for their part, might consider people who give speeches on street corners as brave political activists who demand attention, or as aggressive outsiders and crackpots who should not be taken seriously. Or they might consider themselves 'open-minded', and therefore prepared to pay some attention to the speaker before passing a judgement one way or the other. In the event of a street-corner speech to passers-by, any of these expectations might be confirmed or confounded, and this particular experience might have consequences for how both the speaker and the passers-by approach future instances of street-corner speech-making. So, instances of public address among people co-present in a given site are not somehow unmediated because they are embodied, rather they are mediated by prior expectations and experiences.

A second configuration of bodily presence can be illustrated through the examples of a poster on the street, or a notice left in a bookshop window. In these instances of public address, the bodies of addressers and addressees are physically present in the same site, but at different times. Such non-instantaneous forms of public address are facilitated by bodies combining with media – individuals, organizations, institutions, brands and others have mobilized bodies to deposit a text which is available to those who pass through over a period of time. Here we have a combination of embodiment and mediatization in action. This combination further demonstrates that embodied public address in urban venues is not a matter of static occupation; it can also be a matter of mobile circulation. While some commentators have rightly expressed concern about the political implications of city spaces being made into circulatory spaces (see for example da Landa 1986; Sennett 1978; Rancière 2001), some kinds of circulation also present opportunities for different forms of public address.

Third, texts in a venue of public address might also reach wider audiences through their subsequent mediatization, which allows the further circulation of texts beyond those who are physically present to witness them. Another way to put this is to note that urban sites are not necessarily alternatives to other venues of public address such as print or television; they might also be put to work in a range of combinations. Consider the lamppost sticker on our street corner. As a form of public address, the sticker is by no means limited to capturing the attention of people who physically move through the street corner at some time after the sticker has been applied and before it is removed or deteriorates. A photograph of the sticker might appear in a book devoted to stickers (see for example Dorrian, Recchia et al. 2002), and/or on a website such as www.stickernation.com, and/or in the presentation of an academic giving a seminar about public address and the city. CCTV footage which captured the sticker being applied in the depths of the night might appear on a television news item discussing the ongoing difficulties faced by authorities trying to clear the city of graffiti and other unauthorized forms of writing. And of course, the sticker itself might seek to draw attention to an internet address, a book or some other product. The potential value of a given venue of public address, then, might in fact be enhanced by the possibilities it affords for mediatization. Indeed, the staging of actions by activists in some venues is often calculated to achieve media coverage, even if such coverage can be difficult to direct (Scalmer 2001). Given such potential combinations, it seems to me that any attempt to attach some special value to sites in the city *in opposition to* other mediatized venues of public address such as the internet (see for example Mitchell 2003: 145) is highly problematic.[6]

So, the use of sites in the city as venues of public address can involve different configurations of embodiment, alongside a variety of prior mediations and subsequent mediatizations – the three examples above illustrate this claim while by no means exhausting the possible configurations and combinations. This point has a range of further implications for our understanding of the use of sites in the city as venues of public address. Significantly, universal accessibility to a given site is neither required, nor even desired, in order for it to be put it to work as a venue of public address. Any assessment about the usefulness of a site as a venue of public address cannot be made with reference to any single criterion such as whether it is 'open to all'. This is not to say that different sites all have the same potential to act as venues of public address regardless of their accessibility. Rather, it is to say that the merits of these different sites as venues of public address depend not only on their particular characteristics (such as accessibility) but also the point of view we adopt in making any kind of assessment. Any assessment of the usefulness of a given site to serve as a venue of public address can only be made from the vantage point of a particular public – as such, our assessment of sites will depend on what it is we might want to say, to whom we might want to say it, how we might want to say it, the circumstances in which we might want to connect with others, and our ability to imagine and exploit the opportunities afforded by a given site. The performance of some forms of public address may indeed require or desire a venue which is potentially 'open to all' – a point to which I will return later in this chapter. But some forms of public address actually rely on the limited and particular accessibility of the sites they mobilize. If addressing a public is a matter of imagining and 'finding' a particular audience, then the accessibility of a site to that particular audience assumes more importance than the question of whether it is open to all in a more abstract sense. The very fact that even the strongest advocates of universally accessible urban 'public spaces' choose to write articles and books to outline their case illustrates this point. Those who want to address an academic public on matters of urban theory and politics are unlikely to restrict their activities to street-corner speech-making – their imagined public is elsewhere, and they might also imagine a rather disinterested response to ruminations on Habermas and Sennett from most passers-by.

So, calculations about the usefulness of a particular site as a venue for public address are made with reference to the specific form and content of the address. This point brings us back to our earlier discussion of publics and public address. Putting a site in the city to work as a venue of public address is fundamentally caught up in the queer circularity which Warner (2002) describes – it is an act of the imagination. In contemplating an act of public address – in attempting to connect with others by addressing them or paying attention to them as members of a public – one makes calculations about where those others might be, and what opportunities that 'where' can afford. To address a public, then, not only must we imagine the existence of a public, we must also imagine the existence of venues where address to that particular public can take place. And part of the process of imagining the possibilities of a particular place is the identification of the possible constraints as well as the possible affordances of that place. In other words, our imagination about the possibilities of a particular site is informed by our expectations and aspirations about the kinds of behaviour that are 'in place' or 'out of place'. As such, the public social imaginary is a spatial social imaginary, and spatial imaginations themselves help to *produce* urban space (Lefebvre 1991). What is more, our imaginations of the possibilities of urban sites are increasingly caught up in the transnational circulation of culture, commodities, political ideologies and public forms (Appadurai 1990; Smith 2001; LiPuma and Koelble 2005). This adds a further dimension to the circularity of public address and its relationship to the city – different sites are not only *venues of public address*, but they also become the *object of public address* when their possibilities and limitations are put into question. I now want to explore this dimension of public address in more detail.

2.   Urban places as objects of public address

The very existence of places in cities with distinct 'identities' reflects the attachment of different expectations about what is 'normal' to different locations. Such norms are fundamentally spatial and temporal – the very notion that an action might be 'out of place' in a given spatio-temporal context is an expression of 'expectations about behaviour that relate a position in a social structure to actions in space' (Cresswell 1996: 3). The norms of 'place' are sustained by a range of measures which are deployed to establish limits on the actions of bodies and work to produce bodily capacities which are knowable, predictable, and thereby governable. Of course, such expectations are neither essential to places nor immune to change over time – and they are not always sustained. Places and their identities are in a permanent state of production, and the systems that produce (and re-produce) them are always vulnerable to some degree. The instruments, techniques and procedures mobilized to discipline and govern are always susceptible to resistance and counter-measures, which provoke constant adjustments and improvisations (Hindess 1996). In other words, efforts to produce fixed place identities 'may be totalizing *projects*, but they are not totalizations' (Amin and Thrift 2002: 108). As a consequence:

we must also ensure that we keep a vision of cities with all the uncertainties and risks left in, and especially the recognition that the cities' inhabitants get the chance to redefine, though rarely on their own terms, what it is to be ordered about and interrogated by these systems (Amin and Thrift 2002: 129).

The nature of the expectations attached to different places will have profound consequences for their capacity to be put to work as venues of public address. These expectations act upon our imaginations of what might be possible in given sites. In such circumstances, cities' inhabitants have a variety of options for negotiating and shifting the possibilities of particular places. They may seek to redefine the possibilities for action afforded by a particular place, by making that place the *object* (as distinct from the venue) of public address. This is a second urban dimension of public address, which is concerned with the circulation of contested representations of urban places. Attempts to make socio-spatial norms the object of public address and action might target any combination of institutions of state, economy and civil society that have influence in establishing the identity of places (Emirbayer and Sheller 1999). Such efforts to publicly contest and re-shape the expectations sustained in a particular place are not conducted exclusively *within* the place in question. Rather, those involved in such contests might mobilize a range of different venues of public address to make their case. So, for example, different venues are mobilized to *write* graffiti and to *write about* the regulation of graffiti in those spaces (as we will see in Chapter 5).

However, public action that seeks to re-shape the expectations sustained in a particular place is not the only option for groups who want to expand the possibilities of place beyond existing normative expectations. Groups may also seek to create new possibilities for (counter)public forms of address in a particular place by evading and eluding governance techniques and technologies. In doing so, they may make space for a counterpublic scene of circulation by *avoiding* (rather than pursuing) public action that contests existing socio-spatial norms.

The distinction between the forms of action I have described here follows Michel de Certeau's (1984) distinction between *strategic* and *tactical* spatial practices. In his discussion of the practice of everyday life, de Certeau explored and conceptualized the different 'kinds of action [that] are possible once people historically have been marginalized by a specific regime of "place"' (Morris 1992: 28). He distinguished between *strategic* actions which attempt to (re-)shape the normative expectations and disciplining mechanisms of a given place, and *tactical* actions which ostensibly leave such expectations and mechanisms in place but 'conform to them only in order to evade them' (Certeau 1984: xiv). Both strategic and tactical operations can be deployed in order to re-imagine the possibilities for different forms of public address sustained in the face of normative place identities: 'strategies for the control of space, the definition of boundaries and exteriority; tactics for moving through spaces, (in)visibly, (un)noticed . . .' (Pile 1997: 23).

To make this distinction is not to argue that any groups are restricted to *either* strategic *or* tactical operations. Indeed, groups might shift between 'territorially claiming' and 'temporarily occupying' in different situations (Morris 1992: 28), and we certainly should be wary of de Certeau's own tendency to romanticize tactic over strategy (Ruddick 1996). However, particularly important for my purposes is the notion that strategic and tactical operations have a fundamentally different relationship to public address and representation. When people act strategically in seeking to alter a particular regime of place, they deploy public address in order to shift the possibilities for public address in that place. On the other hand, in pursuing tactical rather than strategic options, a counterpublic 'insinuates itself into the other's place' (Certeau 1984: xix). In shifting between these two kinds of action, counterpublics also make a shift in relation to the publicity of their actions. It is therefore *not* the case that dominant publics and counterpublics are exclusively products of either strategic or tactical action. In particular, when participants in counterpublics attempt to open the possibilities for non-normative forms of public address in a particular place, they confront a set of tensions between:

> on the one hand, the tentative moves, pragmatic ruses, and successive *tactics* that mark the stages of practical investigation and, on the other hand, the *strategic* representations offered to the public as the product of these operations (Certeau 1984: xxiii, original emphasis).

While tactical appropriations imply distinctive forms of invisibility and evasion, the strategic circulation of public representations implies distinctive forms of visibility and confrontation.

This discussion of the tensions between strategic and tactical spatial practices sheds further light on the confusions which characterize the public space debates discussed in Chapter 1. The *idea* of public space, as some of its advocates noted, has been powerfully deployed in struggles over the socio-spatial norms of particular places (see for example Deutsche 1996; Mitchell 2003). That is to say, some model of public space may be publicly circulated by those engaged in strategic efforts to establish a set of normative expectations in a particular space. The ideological mapping of public and private onto particular places can be seen as an indicator of the success of some of these efforts. Historically, these efforts have included attempts to define some spaces as 'private' so as to shelter or banish some practices from collective or public discussion, and attempts to define other spaces as 'public' so as to exclude or curtail some activities on the grounds that they are properly restricted to the 'private' realm. The power of these norms about what is appropriate 'in public' to particular places can produce strong reactions – excitement, disgust and legally sanctioned physical interventions among them. As noted in Chapter 1, moving from 'public' to 'private' can be 'experienced as crossing a barrier or making a transition' (Warner 2002: 26). Indeed, liberal forms of publicness are 'violent' precisely through the operation of generalized categories which work against particularity:

> The inner violence of these principles, including the principle of the public sphere, is rooted in the fact that the main struggle must be waged against all particularities . . . This is the source of the compulsive way in which criteria such as definitions, subsumptions, and categorizations are used to circumscribe the public sphere (Negt and Kluge 1993: 10).

So, it may be correct in one sense to observe that regional conceptions of public and private are ideological, because 'the real social experiences of human beings, produced in everyday life and work, cut across such divisions' (Negt and Kluge 1993: xliii). Indeed, my discussion of urban sites as venues for public address above supports this position up to a point. However, while territorial conceptions of 'public space' might fail to describe accurately the fluid and contested geographies of public address as I understand them, we cannot ignore the role played by these *ideas* of 'public space' in struggles to make the possible uses of particular places objects of public debate. The work performed by the concept 'public space' is not simply mimetic. Our evaluation of the usefulness of 'public space' in understanding struggles over the conditions of public address must consider its normative (and counter-normative) as well as its analytical applications.[7] And this is where the distinction between strategic and tactical spatial practices can be particularly helpful.

Those involved in struggles to make space for counterpublic practices might work strategically to change norms within existing ideological territorializations of 'public' and 'private', seeking to normalize their practices with reference to some model of 'public space'. But in some instances, this strategy may in fact not be the most useful or realistic option open to those engaged in efforts to secure the space for counter-normative forms of public address and sociability. Instead, they may work to change the very territorialization of public and private, seeking to shift the location of the boundary between the two. Or, they might work to evade norms entirely, tactically manipulating opportunities rather than strategically mobilizing regimes of place. Here, existing normalizations of public space are useful only to the extent that they can offer cover for counterpublic tactical appropriations of space, and people may well wish to avoid public engagements concerning norms of 'public space'. The particular tensions and difficulties generated by these different approaches, it seems to me, structure the very production of counterpublic spheres. The question of the usefulness of representations of 'public space' in public debates over the identity of different places is not a question that can be settled in advance – rather, it must be made a matter of empirical investigation in assessing the efforts of particular counterpublics to make their own public urban geographies. We can usefully consider the tensions, then, between strategic efforts to make the norms of a particular place the object of public address and tactical practices which explore alternative possibilities without making a public claim for their normative incorporation.

3. 'The city' as 'the public'

When the norms of public address and sociability sustained in different urban places become the object of public debate, sometimes contributions to these debates are legitimated with reference to the interests of 'the city'. That is, some individuals, groups or institutions may claim to be acting in the best interests of, or even on behalf of, 'the city'. When it is invoked in this sense, 'the city' serves as a proxy for some notion of 'the public'. It refers to a social totality, a collective subject comprised of those who are 'Lagosians', 'Londoners', 'New Yorkers', 'Sydneysiders', etc. This notion of 'the city' as 'the public' is the third urban dimension of public address which I want to consider.

Earlier in this chapter, we considered the argument that concepts of 'the public' served as the imaginary convergence point for the interaction of multiple publics, an idea that could be exploited by the interests contained within each sphere (Negt and Kluge 1993; Warner 2002). Here, '*the* public' is meant to represent the social totality, and those who claim to act in the public interest are thereby claiming that their own particular interests and values are universal. Visions of 'the city' as 'the public' serve a similar purpose. That is, when contests over the norms of different places are conducted with reference to the interests of 'the city', 'the city' functions as an imaginary backdrop for the interaction of different groups, and as such it is an idea that can be exploited in pursuit of particular interests. So, when people claim to be acting in the best interests of, or on behalf of, 'the city', they are effectively positioning their own particular interests and values as universal interests and values.

Like 'the public', 'the city' does not simply *exist* as a collective subject with a stable identity and interests. Rather, visions of 'the city' as 'the public' embody the same chicken-and-egg circularity that is characteristic of publics more generally. 'The city' (as 'the public', the social totality) is performed into existence by the actions of those who act on the basis of its existence – it is a social imaginary. We could take any city to illustrate this process; let's take New York City. In some contexts, a variety of actors may conceive of themselves as 'New Yorkers', and/or address themselves to the collective subject of 'New York'. Indeed, some institutions and groups, such as the 'City of New York', the 'New York Times', the 'New York Yankees', 'New York University', the 'Men in Kilts of New York', and so on, seek to establish some connection to 'the city' in their very name as well as in their deeds.[8] But the 'New Yorks' invoked and addressed by such a wide variety of actors in a range of contexts are not necessarily the same. While the New Yorks performed into being here share the same name, and they may intersect or overlap, they might also be radically different and incongruent with different kinds of boundaries and connections to other places.[9] 'New York', then, has no essential unity, coherence or totality. The very notion that New York (or Sydney, or Lagos, etc.) is a collective subject is premised on 'practices of stabilization' which conceive of actions by a dizzying array of actors as 'parts' which add up to a 'whole' (LiPuma and Koelble 2005).

So, if 'the city' does not simply exist as a stable entity but rather is a social imaginary, this has significant implications for our understanding of contests over the expectations sustained in different urban places. In drawing attention to the imaginary dimensions of conceptions of 'the city' as a collective subject, my point is not to argue simply that 'the city' (as New York, Sydney, London, Lagos, etc.) is not *real*. Indeed, quite the contrary. As LiPuma and Koelble (2005: 175) note, 'it is necessary to apprehend the city as a totality because others think of it that way. They act, desire, and conceptualize the future on that basis'. As such, these urban imaginaries have quite real effects, to which we must be attentive. In analysing the urban dimensions of struggles over the conditions of public address, we must therefore critically interrogate the urban imaginaries that are mobilized in efforts to shape the possibilities of particular places for different forms of public address. Do particular urban imaginaries, with their associated visions of what is in the best interests of the city/public, privilege some interests over others by making them appear 'universal' rather than 'particular' or 'parochial' or 'private'? Do the 'shapes' of these different imagined cities deprive anyone of 'symbolic enrolment in the city' (Rancière 1999: 23)? In other words, just as critical theorists of the public sphere have drawn attention to the particularity of the interests and values advanced by those who claim to be acting on behalf of *the* public, so too we must ask: which particular interests are advanced when some individual, group or institution claims to be acting on behalf of 'the city'?

In asking critical questions about the nature of urban imaginaries which mobilized in contests over what is 'normal' and what is 'out of place', we can identify two kinds of urban imaginary which have come under attack by critical urban theorists concerned with the heterogeneity of the urban public sphere. First, a range of criticisms have been levelled at the notion that the city can be constituted as a collective subject on the basis of the *shared values* of its inhabitants. Second, criticism has been directed at urban imaginaries premised on the notion that the inhabitants of the city have a set of pre-political *shared interests*. Against these visions of 'the city', some critical urban theorists have sought to conceptualize 'the city' in a new way that recognizes its fundamental heterogeneity. Their efforts have been informed by the work of those critical social theorists discussed earlier in this chapter who sought to articulate the contours of a wider public sphere which is premised on the interaction of multiple publics. This third vision of 'the city' is premised on the notion that the *shared fates* of urban inhabitants are the only thing that binds them together as a collective subject. I now want to survey these debates briefly, in order to identify some of the lines of inquiry which might be useful for our subsequent analysis of the different urban imaginaries which are mobilized by participants in struggles over the urban dimensions of public address. In discussing these different visions of 'the city', I want to focus on how they conceptualize the figure of *the stranger* and his/her place in the city/public. The figure of 'the stranger' plays a key role in the urban imaginary. As LiPuma and Koelble (2005: 156) note, 'the urban imaginary mediates the generative schemes by which strangers live collectively, being thus able to think of themselves in certain contexts as, for example, a public'. The question of how strangers imagine themselves as a city/public with shared horizons is therefore simultaneously a question of whether they can produce and participate in some kind of wider public sphere or public realm.

First, then, let us consider the notion that shared horizons among urban inhabitants can only exist as the product of shared values and ways of life. From this perspective, in the good city, strangers overcome their estrangement by creating some 'sense of community'. Recently, such a vision of the city has been particularly influential in urban design circles, and the physical layout of cities is seen to play an essential role in the creation of 'community'. Here, the estrangement of large populations in cities is to be combated by encouraging the formation of 'urban villages' where people can once again get to know their neighbours, thus producing a 'new urbanism' (see for example Katz 1994). This 'new urbanism' is clearly influenced by the work of Louis Wirth and other urban sociologists of the 'Chicago School', who identified urbanism as a way of life characterized by the breakdown of communal bonds. The disruption of traditional forms of socialization associated with village life was seen to be giving way to the differentiation of urban populations and creating opportunities for deviance. The key priority for urban politics was thought to be re-establishing new communal bonds to overcome the estrangement that produced differentiation and deviance (Katznelson 1992). Like the Chicago School urban sociologists, the 'new urbanism' seeks to build a shared horizon among urbanites by giving 'the reassuring impression of an integrated society, united in facing up to its "common problems"' (Castells 1979: 85). The drive towards shared values and community is a problematic basis for any city/public:

> If community is a positive norm, that is, if existing together with others in relations of mutual understanding and reciprocity is the goal, then it is understandable that we exclude or avoid those with whom we do not or cannot identify (Young 1990: 235).

According to critics, the new urbanists' vision of the city as a community is therefore likely to be mobilized in support of exclusionary measures designed to deal with those who cause problems because they don't 'belong' (Harvey 2000).

A second powerful vision of the 'city' as a collective subject is premised on quite a different understanding of the common interests shared by strangers in the city. Here, urban inhabitants with different values are to be treated as equals in a 'public city' which recognizes their *shared interests* as citizens (Watson 2002). The inhabitants of a city are imagined to have a set of collective interests and needs that can be addressed through state provision of universally accessible services and infrastructure. Access to the public city is a matter of citizenship, to be facilitated regardless of social and economic status. The public city is contrasted with the capitalist city in which access to goods and services is procured 'privately' through the market. Through state action in the public interest, the unequal outcomes of market exchange can be ameliorated. This vision of the city has been particularly influential in some planning circles over the course of the twentieth century. In the public city, according to its advocates:

> public institutions and public space were seen as providing symbols for a sense of community and civic morality that overrode the individual commercial interests of the inhabitants of the city (Davison 1994: 4).

This vision of the city continues to be influential in contemporary public space debates in urban studies, where the commercialization of public space is said to undermine the genuinely inclusive public city because it supplants citizenship with consumer status in determining access. Peter Spearritt (2000: 82), for example, suggests that in 'real' public space, 'you are regarded as a citizen, as a legitimate member of society even if you don't intend to make a purchase of a product or service'. But increasingly, says Spearritt, a range of developments have established the consumer rather than the citizen as the ideal user of public space:

> If your access to particular public spaces depends on both your willingness and your ability to pay, then those spaces cease to be public. They have effectively been added to the private domain (Spearritt 2000: 96).

Unlike visions of 'the city' as a community, visions of the 'public city' usefully draw attention to the potential for unequal outcomes associated with the unchecked application of market forces in the determination of resource allocation and distribution. But this urban imaginary is not without its limitations. In particular, the identification of the 'public interest' with the state has too often rendered it the preserve of expert planners, who are supposed to possess the technical capacity to somehow rise above the competing self-interests of urban inhabitants in order to determine the common good. But planners have in many instances been more concerned with rationality than democracy, their interventions serving to discipline the city to an imposed order. A colleague of Robert Moses, a post-war planner responsible for many large-scale public planning projects in New York City, once remarked that 'He loves the public, but not as people' (quoted in Sennett 1994: 363). The actions of state agencies may in fact be more concerned with administration and governance than agonistic democratic exchange across difference. Hannah Arendt (1958) argued that state concern with administration and juridification contributed to the rise of the *social*, a realm of life quite distinct from the *public* realm in which people appear before one another as equals and unique individuals come together to debate and reflect upon their common interests. It is important to distinguish between the state and the public, and to be wary of conceptions of the public interest that fail to acknowledge the fundamental diversity of urban inhabitants.

In response to the limitations of the two urban imaginaries just discussed, a third vision of 'the city' is gaining increasing influence in critical urban theory: the city as 'cosmopolis'. This is the urban imaginary that is most closely aligned with the visions of a heterogeneous public sphere discussed earlier in this chapter. In the cosmopolis, strangeness is not a condition to be overcome or set aside in the public interest – rather, estrangement is accepted as the fundamental condition of urban life, and it is to be drawn upon as a resource. In this vision of the city, a shared horizon which does not seek to erase estrangement and heterogeneity can be constructed if urban inhabitants are willing to find ways to deal with their *shared fates*. It is these shared fates, rather than shared identities or interests, which are said to bind city dwellers together into a polity (Young 1990: 238). If the problems facing this polity are to be addressed, this 'does not necessarily mean the return of the outmoded concept of "the public interest", but it does demand the creation of a civic culture from the inter-actions of multiple publics' (Sandercock 1998: 187). The creation of this cosmopolitan civic culture is a delicate balancing act. On the one hand, participation in the city/public should not be premised on the 'adoption of a general point of view that leaves behind particular affiliations, feelings, commitments, and desires' (Young 1990: 105) as in visions of the city/public constituted by shared values or interests. On the other hand, participation in a cosmopolitan civic culture does require 'an ability to take some distance from one's immediate impulses, intuitions, desires, and interests in order to consider their relation to the demands of others, their consequences if acted upon, and so on' (Young 1990: 105). In other words, urban inhabitants are asked to be 'reasonable' (Young 2000: 24–5; Deutsche 1999). In this balancing act, then, participants in the cos-mopolitan city/public must address their shared fate by acknowledging that they are strangers to each other *and* that they are 'strangers to themselves' (Kristeva 1991). A cosmopolitan public dialogue about shared problems is premised on a willingness not to fix our ideas of each other and ourselves in advance of participation in that dialogue. In Kian Tajbakhsh's (2001: 183) terms, the 'promise of the city' is 'the freedom to glimpse our own hybridity, our own contingency'. This can only emerge from a 'cosmopoli-tan ethic' of 'openness to the other proceeding from the recognition of the stranger within us' (2001: 9).

This cosmopolitan urban imaginary has the strong advantage of recog-nizing and addressing the multiple publics which are likely to constitute the city as a public. Advocates of the cosmopolitan vision of the city/public hope to avoid the exclusions of other visions of the city/public by re-imagining a shared horizon which is not premised on shared values or interests. But like all publics, the cosmopolitan city/public cannot be imag-ined without some horizon which sets limits on its scope. The notion that strangers in the city share a fate is no exception. This notion establishes criteria for participation which have exclusionary consequences. City-dwellers are said to be bound together, and they must then choose how to respond to this condition. This togetherness is said to demand 'openness' and a capacity to be 'reasonable'. Here, while the values or interests of par-ticipants in the city/public are not established in advance of their partici-pation in the city/public, their identity *as city-dwellers* is. This is inescapable because of the circularity of public address, but we must not be blind to its consequences. The demand for people to be 'reasonable' is an attempt to establish criteria which sets limits on the city/public. These limits may be preferable to those associated with other visions of the city/public, but they are not without potential exclusions. As I have asked elsewhere (Iveson 2006a):

> if graffiti writers refuse to put their identities at risk by engaging in a wider dialogue over urban aesthetics with 'outsiders' to their subculture, are they being unreasonable? Should they be lured or forced into such a dialogue, because living together in the city demands it? If the homeless and the addicted fail to participate in a debate about the norms which govern street contacts and begging, are they being unreasonable? Should they be 'educated' or 'empowered' to participate in such a debate, because living together in the city demands it?

Even a vision of the city/public as a 'being together of strangers' sets up potentially powerful dynamics because it inevitably privileges the city as the shared horizon of the public, as if this was not a product of politics but rather a matter of fact. But as Rancière (2001: para 4) has observed: 'What is proper to politics is thus lost at the outset if politics is thought of as a specific way of living. Politics cannot be defined on the basis of any pre-existing subject.' If one conception of the 'city' is preferable to another, why is a conception of the public as a 'city' preferable to other conceptions? Healey's (2002: 1779) advocacy of constructing the 'city' as a collective discursive resource clearly posits the answer to this question in terms of enhancing governance capacities:

> the strategic capacity to imagine the city, and to mobilize wide-ranging attention and action around discussion about such conceptions, has the potential to become a key element of the institutional infrastructure of urban governance.

As such, we must always be attuned to the particular interests and ways of life that are privileged by different ways of imagining 'the city' as a collective subject. And rather than demanding that urban inhabitants be open to encounters with 'strangers', we need to learn more about the circumstances in which particular people have taken the risks associated with these strange encounters and transformed their cities in the process.

So, what are the implications of this discussion of competing urban imaginaries for our inquiry into the urban dimensions of public address? Most significantly, this discussion suggests that we should seek to identify the exclusions that are sustained by visions of 'the city' mobilized in contests over the spatio-temporal expectations attached to particular places. Any vision of 'the city' as 'the public' will inevitably fail to live up to its claim to represent the social totality, because like all publics, the city's limits are indefinite but not infinite. However, in identifying these exclusions, my aim is not to derive a normative vision of 'the good city' which can overcome all exclusions. Nor do I intend to measure the actions of those engaged in struggles over the making of publics against some normative model of the 'reasonable' citizen derived from such a vision of the good city. Rather, the normative project of this book is more pragmatic. I want to critically analyse the ways in which different people have made space for non-liberal forms of public address in and through the urban. As such, I want to trace the different ways in which those engaged in struggles over the making of publics imagine the strangers on their horizon. I am neither for nor against 'the city' as the privileged terrain of an ideal form of political community, but I am certainly interested in understanding its effects in both opening up and closing down 'partitions of the perceptible'.

Investigating Public Urban Geographies

We now have a clear framework for analysing the urban geographies of struggles over the conditions of public address – a framework which is sensitive to the multidimensionality of both publicness and the city. To recap: I have characterized the public spheres which sustain different forms of public address as a distinct form of social imaginary. I then argued that public social imaginaries are also spatial imaginaries, and pointed to three related but distinct urban dimensions of public spatial imaginaries. Struggles to make space for different forms of public address involve:

- efforts to use different urban sites as venues of indefinite address;
- efforts to re-imagine the possibilities of different urban sites by evading socio-spatial norms and/or making them the object of representations and public debate; and
- efforts to legitimate public representations and claims about the possibilities of different urban sites with reference to different visions of the city as a collective subject, a common horizon of existence shared with other urban inhabitants in a particular way.

Analysis of the urban dimensions of struggles over the making of publics will require us to explore the interactions between efforts on all three of these fronts.

With this framework in place, we are in a position to begin our analysis of the urban dimensions of struggles over the conditions of public address. The case studies to follow focus on struggles to produce and sustain different forms of publicness through the city. In each chapter, I chart the actions of those involved in these struggles across the three urban dimensions of public-making outlined above. The first three case studies examine efforts to sustain and re-shape opportunities for particular non-normative forms of public address in shared spaces of a particular city. In Chapter 3, I trace struggles over the use of the Parliamentary precinct in Canberra for protestatory public address. The chapter examines the ways in which the conduct and regulation of protest became enmeshed in wider debates about the identity of the precinct in relation to Canberra's role as the National Capital. Chapter 4 moves to Melbourne, and explores efforts to re-shape the context in which men cruise 'beats' (places such as parks, rest rooms, etc. where men seek sexual encounters with other men). As cruising became an object of concern for community-based health workers and gay and lesbian rights campaigners, their interventions sought to make a 'proper place' for cruising in the city by contesting hetero-normative policing of beat spaces, with ambiguous results for the sexual counterpublic sphere produced by cruising men. In Chapter 5, I look at the ongoing mutation of graffiti-writing in Sydney. This chapter focuses on the nature of efforts to sustain and contain different forms of graffiti-writing, examining the complex ways in which these efforts have impacted upon the counterpublic sphere which enables the circulation of graffiti as public texts.

Having explored struggles to sustain non-normative forms of public address in these three chapters, the next two case studies examine efforts to secure space on behalf of particular normative forms of public sociability through exclusionary means. In Chapter 6, I look into the re-making of 'public space' in Perth. Here, I carefully trace the way a loose coalition of actors came together to promote a new, 'neo-liberal' vision of Perth as a 'Capital City' for Western Australia, and I examine their efforts to enact this vision by producing a new kind of 'public space' for the city centre. In the process, certain people (particularly indigenous young people) were positioned as outside of 'the city' – through efforts to deny them access to its physical spaces and to the political spaces of public debate and policy deliberation through which city visions were formulated and potentially contested. The chapter points to the contingency of these efforts, and it also considers a range of efforts to redirect neo-liberal discourses and techniques of urban governance in order to sustain other forms of public sociability in the city. Chapter 7 examines a campaign to exclude men from a public swimming pool reserved exclusively for women and children in Sydney, after one man made a legal complaint of discrimination on the grounds of sex when he was denied access to the pool. The chapter charts the spaces and arguments through which this campaign was fought, examining in particular the ways in which the campaign politically framed its key claims in the wider public sphere. Taken together, these two chapters demonstrate the importance of considering the interaction between the three urban dimensions of public-making when exploring struggles over the conditions in which people come together as a public. While both sites are exclusionary, I reveal important differences between these forms of exclusion by exploring the kinds of debate which sustained exclusion in these sites and the visions of the city which were mobilized in these debates.

In each of these case studies, then, I aim to respond to Warner's (2002: 61) call for 'both concrete and theoretical understandings of the conditions that currently mediate the transformative and creative work of counterpublics'. The case studies are intended to be illustrative rather than representative – I make no claim that the processes of urban restructuring and struggle discussed in the chapters to follow are indicative of trends being replicated elsewhere (it does not all come together in Perth, or in any other city for that matter!). Nonetheless, I do hope that readers interested in cities elsewhere will find interesting points of comparison. Protest, graffiti, cruising, neo-liberal urban restructuring and even women-only recreational facilities are by no means unique to the Australian cities under investigation here. Further, it is my strong hope that the approach I have taken to analysing these various struggles might usefully be adapted to inform the analysis of public-making in other urban contexts. As such, in the final chapter, I seek to draw out some points of comparison across the case studies in order to offer some more general reflections on the ways in which public urban geographies are produced.

Chapter Three

Making a Claim: the Regulation of Protest at Parliament House, Canberra

Marches, assemblies, rallies, pickets, 'zaps' – all of these forms of public address make use of different urban sites as venues for political claim-making. The protestor, by mobilizing some place in the city in order to make a political claim, is engaged in a form of urban public address which has occupied a central place in many accounts of the city's contribution to democratic citizenship. For advocates of urban 'public space', the importance of places where citizens can gather to engage in protest has been a recurrent, if not dominant, theme. In their accounts, when all other channels of participation and dissent have failed, cities ought to provide 'public spaces' which serve as 'spaces for representation' where those who have a claim to make as citizens can congregate in order to be seen by other citizens, thereby drawing attention to injustice and oppression (see for example Mitchell 2003). And while some have claimed that such contentious gatherings in urban spaces have become increasingly anachronistic in the digital age (see Chapter 1), cities around the world continue to sustain a tremendous diversity of protest actions. Indeed, the fact that events associated with the growth of anti-capitalist politics in recent years are often referred to simply by the names of cities (such as 'Seattle', 'Prague', 'Genoa', and 'Porto Alegre') attests to both the continued significance of contentious gatherings and assemblies in cities for contemporary politics and the importance of emergent global media-scapes in producing this significance.

This chapter investigates how the possibilities for protest in a particular place – Parliament House in Canberra, Australia's capital city – have been identified, exploited, debated and regulated since the building was officially opened in 1988. Parliament House in Canberra makes an excellent place from which to study the urban dimensions of political protest, for a number of reasons. Parliamentary precincts located in capital cities have acquired a heightened significance for protestatory public address in a variety of national and state contexts. This is associated with the general parliamentarization of popular politics that began to gather pace in Western countries in particular in the eighteenth and nineteenth centuries (Tilly 1997). While actions across a range of urban spaces have been staged in order to put claims to parliaments, gathering at the Parliament itself has been an important way for social movement actors to have addressed Parliamentarians and the wider public they are supposed to represent. Protest at the national parliament building in Canberra is particularly ripe for research. At the time of writing, the building is less than twenty years old. The massive efforts to invest the Parliamentary precinct in Canberra with national significance and post-colonial identity are still relatively young. Debates over how the place can be used as a venue for political claim-making have been central to these place-making efforts. The opportunities and constraints that the new building affords for contentious gatherings have been a problem for both Parliamentary authorities and protestors alike, as both seek to respond to exploratory interventions from each other.

The chapter begins with a brief history of the making of Parliament House and its surrounding precinct, charting the efforts of the designers to craft a place with a quite particular identity in relation to the Australian nation-state. This sets the scene for a discussion of some of the significant protest events that took place at the Parliament in its early years, looking at how these events both mobilized and challenged the precinct's identity in order to stage different forms of public address. These events eventually provoked a Parliamentary Inquiry into the 'Right to Protest' at the Parliament. Next, I consider the deliberations of this Inquiry, which resulted in the consolidation of a set of legally enforceable *Guidelines* for protest. These *Guidelines* quite clearly establish a set of time-space expectations for what we might call 'proper' forms of protestatory public address in the Parliamentary precinct. In subsequent years, while some protestors have complained that these *Guidelines* curtail their rights to protest, others have continued to explore ways around the *Guidelines* which have tactically appropriated existing opportunities in pursuit of political representation. Paradoxically, perhaps, these latter efforts suggest that being 'seen' is increasingly a matter of being 'unseen' in some important respects. Finally, the chapter concludes with some critical reflections on the efforts of Parliamentary authorities to privilege some forms of public address over others.

The Making of a New Parliament House

Designing the new Parliament

The year 1988 marked the 200th anniversary of the British colonization of Australia. Amidst a year of Bicentennial celebrations, Queen Elizabeth and Prince Philip officially opened a new Parliament House on Capital Hill in Canberra on 9 May. Architect Romaldo Giurgola's design for the building had been selected in 1980 following an international design competition to which over 300 entries were submitted. By the time of its completion in 1988, the winning design had cost over AUS$1 billion to realize. Parliamentary functions moved to the new building shortly after its opening, having been conducted for over sixty years in a building that was designed to be temporary. The 'provisional' Parliament House (now known as Old Parliament House) had been overcrowded throughout most of the post-war period, too small to accommodate the growth of the Parliament and unable to facilitate the effective use of new communications technologies increasingly used by both Parliament and the media.

The construction of a new and permanent Parliament House presented the Australian Commonwealth Government with an opportunity to make a grand symbolic statement about national identity and governance. In Canberra, as elsewhere, architecture has been mobilized as part of a wider attempt to 'rebrand' the nation, in relation to both its place in the world of nations and its colonial past (see also McNeill and Tewdwr-Jones 2003; Yeoh 2001). The design brief for potential architects informed them that:

> Parliament House must be more than a functional building. It should become a major national symbol, in the way that the spires of Westminster or Washington's Capitol Dome have become known to people all over the world (Parliament House Construction Authority 1979b: 15).

In pursuing this symbolic project, Giurgola proposed a building that 'would embody the spirit and character of the nation' through a fusion of architecture and art:

> the choice of artists and works of art must be made at a time and in a manner in which their involvement in the project is not at the end of a sequence, but rather is an integral part of the process of development of the interior design (quoted in Commonwealth of Australia 1993: 9).

This philosophy is reflected in the choice of materials used in the building, as well as the choice of artworks, many of which were specially commissioned for the new Parliament House.

But exactly what kind of national identity was to be recognized in the new building? As Warden (1995: 55) has noted:

While the old provisional Parliament House was steeped in the symbols of constitutional monarchy, the new and permanent Parliament House is resonant of a new sense of national identity which may be called post-colonialism.

Themes of multiculturalism and prior occupation of Australia by Aborigines and Torres Strait Islanders take their place alongside more traditional celebrations of the Australian nation in the art collection. The design of the new Parliament House also attempts to be suggestive of a particular liberal-democratic relationship between politicians and the people. Tour guides tell visitors that regardless of where they are in the building, they are never overlooked by Members of Parliament. Most famously, this is achieved by the setting of the building into Capital Hill, which allows visitors to walk up grassed ramps to the building's pinnacle, under an enormous flag mast.[1] Large public galleries also overlook both Parliamentary chambers. According to one of the documents produced to celebrate the opening of the new Parliament House:

From its spacious entrance through wide public rooms and galleries, the new Parliament House welcomes visitors and invites participation in the democratic process . . . The invitation to all Australians to observe the democratic process is a central design feature of the new building (Parliament House Construction Authority 1986: 3, 19).

This symbolic aspect of the design project, however, has not been pursued to the exclusion of other concerns – while the building was to be 'more than functional', it was also to be functional. Indeed, the Parliament House Construction Authority (PHCA), charged with the responsibility of both conducting the design competition and overseeing the construction of Parliament House, was committed to ensuring that the new building would provide the Parliament 'with a highly efficient environment in which to carry out its increasingly complex responsibilities' (Parliament House Construction Authority 1986: 2). It had to be large enough to accommodate both Houses of Parliament, the Executive Government, a wide range of support and ancillary services, the media, and a large volume of visitors. Visitors posed a particular problem. The very scope of the new Parliament House, and its central place in Canberra's symbolic landscape, meant that it was sure to generate huge numbers of tourists. The PHCA (1979b: 9) worried that 'without regard to movement segregation, the high numbers of visitors could cause some problems'. It therefore stipulated that 'all areas of interest to visitors should be grouped and linked by a circulation system discrete from that used by people working in the building' (Parliament House Construction Authority 1979b: 9). At the provisional Parliament House, King's Hall was the main space between the two Parliamentary chambers, and it was open to 'the public'. Journalists frequently staked out this space in order to talk to politicians, and in 1982, it was occupied by striking mineworkers. No such fluid space was to be provided in the new building:

> The circulation space around and between the Chambers at Chamber floor level should not be accessible to tourists. This is essential to maintain security and the efficient operation of the business of the Chambers (Parliament House Construction Authority 1979b: 35).

As the guidelines for the design of the new building were being developed in 1978, concerns about security were heightened after the explosion of a bomb at the Hilton Hotel in Sydney (the venue of the Commonwealth Heads of Government Meeting), which resulted in the death of one council worker and injuries to one police officer. According to the Parliament's Security Controller, speaking in 1998 on the tenth anniversary of Parliament House's opening, the Hilton bombing:

> proved to be a galvanizing factor for security and counter-terrorism policy in Australia. Soon after, the present security organization within Parliament House began to take shape . . . (Taylor 1998: 46).

After the bombing, the Australian Security Intelligence Organization was engaged to provide a full-time security consultant to the PHCA on the security aspects of the building's design and construction. Like the symbols of nationhood, security needs were to be designed into the building as a priority: 'security was to be achieved first through design, second using equipment, and third using staff at security points' (Hume 1996: 8). The PHCA's (1979a: 30) design brief noted that 'although Parliament House is, in a symbolic sense, the manifestation of a democratic way of life, with connotations of openness and freedom, security cannot be ignored'.

Giurgola's winning design took these security concerns into account by segregating different traffic flows and simultaneously creating the impression of openness and interaction for visitors:

> the two main traffic flows, Parliamentary and public, are separated. Parliamentarians enter the Chambers and Halls on the ground floor; visitors move from the Foyer to the first floor, and from there either along the public circulation areas adjacent to the curved walls or past the Halls at gallery level. Glazed pedestrian bridges then lead to the public galleries of the House of

Representatives or Senate Chambers (Parliament House Construction Authority 1986: 17).

The different traffic flows have their own discrete entrances. The main entrance to the building is used by members of 'the public', while separate restricted entrances are provided for Senators, Members of the House of Representatives, and Ministers some distance away from the front of the building. As one security analyst reflected some years later:

> Keeping those visitors within a circulation pattern and maintaining the feeling of free movement would discretely prevent access to areas reserved solely for carrying out the business of Parliament and the Executive (Hume 1996: 8).

So, two related but distinct conceptualizations of 'the public' were mobilized in the design and construction of Parliament House. In one sense, the entire building was intended to be a 'public' building, constructed and operated by the Australian state on behalf of 'the people' (as 'the nation') and housing the operations of their elected representatives. In another sense, significant parts of the building were to be off-limits to actual members of 'the public' who might visit the Parliament. From this perspective, the activities of 'the public' were delimited by those seeking to establish the territorial configuration of the new building. If 'the public' owned the building and was represented in it in a symbolic sense, in a functional sense 'the public' constituted a particular category of user – the 'visitor' – who was to be kept apart from both Parliamentarians and Parliamentary staff. Visitors' movements were to be incorporated into the building but segregated from the operational spaces of the Parliament itself.

The Place of Protest at Parliament House

Even within these designed limits, the possibilities for visitors to use Parliament House as a venue for political protest seemed quite open when the building was still in construction. As one observer noted in 1987 shortly before the building was completed, Giurgola's design seemed to encourage the staging of protest in number of spaces around (and on top of) the building which were to be accessible to visitors:

> Conceived as a Parliament for the people, the new building permits public access to wide forecourts and a grassed roof area which could provide ideal locations for protest meetings (Handley 1987: 27).

However, as we will see, such possibilities were almost immediately fore-closed. The forecourt and the grassed roof area are both now off-limits to protestors. So too are the Ministerial and Members' entrances to Parliament House. While one can visit these spaces, visiting them with a placard or with the intention of handing out leaflets will result in a direction to move to a designated protest area by an officer of the Australian Protective Service. Failure to comply with this direction could result in arrest.

Assigning a proper place for protest has been a key element of the place-making project which has sought to invest Parliament House with its spatial identity. In the design and construction stages, it was always anticipated that the new building would attract protests and demonstrations – protest was not considered to be 'out of place' in the Parliamentary precincts. Indeed, the staging of protest was thought to be confirmation of the sym-bolic status of Parliament House as 'an emblem of Parliament's central place in Australian society' (Parliament House Construction Authority 1986: 3). A Parliamentary report into the right to protest (discussed in more detail shortly) put it this way:

> As Canberra is the seat of the Federal Government and the home of the national Parliament, it is recognized as the place where decisions are made which affect the lives of all Australians. On this basis, Canberra also is the place where people come to express their views on those decisions and other matters of importance to their lives (Joint Standing Committee on the National Capital and External Territories 1997: 2).

Such statements by politicians are not simply evidence of a grudging resignation to the inevitability of protest, rather they are made with a degree of pride and/or self-importance. Nonetheless, Parliamentary authorities have self-consciously sought to define and delimit the ways in which pro-testors can use the Parliamentary precincts as a venue for public address and political claim-making. 'Protestors' – like 'visitors', 'Parliamentarians', 'dignitaries', 'journalists' and 'delivery vehicles' – have been assigned a place at Parliament House in a manner consistent with the principles of func-tional separation and Parliamentary privilege. While protest is allowed, authorities are adamant that it should not disrupt the efficient functioning of the building by impinging on other users and uses, or disrupt the iden-tity of the building as the home of modern Australian Parliamentary democracy.

Defining and enforcing the 'proper place' for protestors has been a con-stant challenge for Parliamentary authorities. The Speaker of the House of Representatives and the President of the Senate (collectively known as the Presiding Officers) have responded to this challenge by issuing the *Guidelines for Protests, Demonstrations and Public Assemblies within the*

Map 3 ~ The Parliamentary Precincts, the authorised assembly area, and Federation Mall

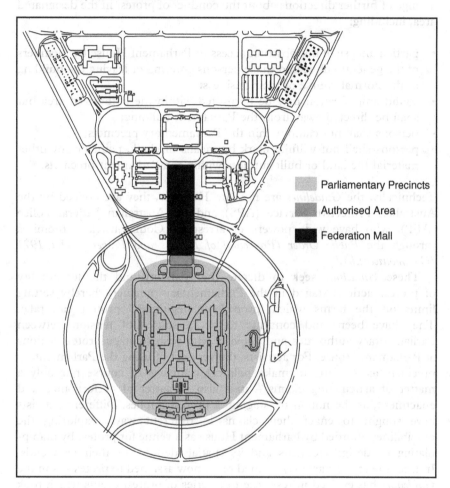

Figure 3.1   Map of Parliamentary Protest Zone, Canberra (source: National Capital Authority *The Right to Protest*)

Parliamentary Precincts (hereafter the *Guidelines*). If one or more persons wishes to 'meet or engage in a public activity for the purpose or on the pretext of making known a grievance or discussing public affairs', they are directed to do so in a designated protest area about 150 metres from the main public entrance, and over twice that distance from entrances used by Parliamentarians (people may be present at these entrances if they are not protesting) (see Figure 3.1). Any structures such as stages must be

approved by the Parliament House Security Controller, and will not be permitted to remain in the designated area overnight. The *Guidelines* contain
a range of further directions about the conduct of protest in the designated
area, including:

- participants shall not obstruct access to Parliament House by members
  of the general public and those persons entering or leaving the building
  in the normal course of their business;
- sound amplification may be permitted within the authorized area but
  shall be directed away from the Parliament building;
- persons shall not camp within the Parliamentary precincts;
- persons shall not wilfully mark by means of chalk, paint or any other
  material the land or built surfaces within the Parliament precincts.

Technically, the *Guidelines* are not law. However, they are policed by the
Australian Protective Service (APS) and the Australian Federal Police
(AFP), who have wide powers of arrest on Commonwealth premises
through the *Public Order (Protection of Persons and Property) Act 1971
(Commonwealth)*.[2]

These *Guidelines* seek to dictate the spatiality and the temporality
of protest actions staged in the Parliamentary precinct, thereby setting
limits on the forms which 'proper' or 'legitimate' protest can take.
They have been, and continue to be, a source of tension between
Parliamentary authorities and activists who wish to stage protest actions
at Parliament House. For activists, the very act of using the Parliamentary
precincts as a venue to make political claims is of course not only a
matter of articulating claims – it is also a matter of performance and
enactment, as the notion of 'staging' a protest implies. Different activists
have sought to enact their claims in different ways, exploring the
possibilities afforded by Parliament House as a venue for protest by manipulating its design elements and its spatial identity for their own ends.
Indeed, the rather narrowly defined place now assigned to protestors in the
Guidelines has evolved in response to a series of protest events which took
place in the years following the opening of the new Parliament House. So
troubling were some of these events to Parliamentary authorities that in
1995 the Commonwealth Government established a Parliamentary Inquiry
into 'the right to legitimately protest on national land'. The deliberations
of this Inquiry provide an excellent window onto the perspectives of
Parliamentary authorities and political activists alike, and were fundamental in establishing and consolidating the current *Guidelines* for protest.
Before considering the Inquiry, I want to offer some snapshots of significant protest events that took place at Parliament House in the preceding
years.

Figure 3.2   Forecourt Mosaic, Parliament House (source: *Parliament House Pictorial Guide*)

Protest Snapshots, 1988–95

Rewriting the meaning of the Forecourt mosaic, 1988 and 1993

Romaldo Giurgola's conception of a building in which art and architecture were fused is perhaps best expressed in the Forecourt mosaic, designed by Aboriginal western desert artist Michael Jagamara Nelson (see Figure 3.2). Nelson's *Possum and Wallaby Dreaming* was chosen from works of art by five western desert artists who had been invited to submit designs. According to the book documenting the Parliamentary art collection:

> The Forecourt mosaic announces the theme of the land in a work which expresses traditional Aboriginal culture in a contemporary form. The mosaic

stands within a pool on an island representing the continent of Australia, occupied by Aboriginal people for thousands of years (Commonwealth of Australia 1993: 13).

On the day of its opening, Aboriginal protestors used this mosaic as a means to protest about their treatment by the Commonwealth Government.

During the official opening ceremony, Michael Jagamara Nelson escorted the Queen, Prince Philip, Prime Minister Bob Hawke and his wife Hazel Hawke to the forecourt, and explained to them the meaning of the mosaic. Earlier in the day at dawn, a group of senior Aboriginal women from Papunya had performed their own ceremony dedicating the mosaic to its Dreaming purpose. While the mosaic has a particular meaning attached to it in official discourse about the new building, it also has a meaning in relation to the Dreaming of Michael Nelson and the Warlpiri people. Vivien Johnson (1997: 91) writes:

As far as Michael Nelson and the dancing women were concerned, their ceremony of dedication had awakened the ancestral power of the ancient symbols and made that area of the forecourt more a sacred site than a work of art – a powerful Aboriginal presence that was permanently embedded on Australia's front door step to remind it of its obligations.

One of the Aboriginal protesters present on the day of the opening tried to attach another meaning to the mosaic – activist and poet Kevin Gilbert held a press conference and informed the assembled journalists that the mosaic had put a 'curse' on white Australia. *The Australian* (10 May 1988: 1) reported that:

An Aboriginal artist, who yesterday proudly showed the Queen the mosaic he designed for the new Parliament House, used it to place an elaborate curse on white Australia, it was claimed yesterday . . . Mr. Kevin Gilbert – a prominent if eccentric black activist and artist – topped off a day of Aboriginal protest against the $1 billion landmark with the startling claim that the curse would haunt the Federal Government until justice was delivered to Aboriginal people.

Gilbert felt that as long as justice had not been achieved, the incorporation of Aboriginal art such as the mosaic in the new building was exploitative:

White Australia has tried to exploit that [the mosaic] as being symbolic of some sort of honourable dealing with Aboriginal people, as a meeting place (Kevin Gilbert, quoted in *The Australian*, 19 May 1988: 1).

The media reports did not include a comment from the artist himself, and noted that Gilbert's claims could well have been a hoax. On 10 May, Nelson held his own press conference, flanked by Aboriginal leaders Charlie Perkins and Kaye Mundine. He refuted Gilbert's claims, saying 'I've done this painting for good purposes, for both white and black Australians' (Johnson 1997: 90). The inclusion of the mosaic in the design of Parliament House not only allowed Gilbert to highlight his case for justice in the media, it also gave Nelson an opportunity to highlight his art and its intentions. But Gilbert's protest raised concerns for Parliamentary authorities about the ways in which the mosaic might be used by protestors. Its status as a work of art was used to argue that protestors should not be allowed to congregate on the forecourt in a manner that might challenge or manipulate the authorized meaning of the mosaic. Future protests were restricted to any area on the other side of Parliament Drive.

However, in 1993 Nelson asserted his own authority as artist to use the mosaic in protest. In 1993 the Keating government proposed a new regulatory regime for Aboriginal land rights claims. On Monday 27 September, a group of about 500 Aboriginal protesters gathered on the forecourt of Parliament House to protest against the legislation, and to press for more time to 'get it right'. Among the protesters was Nelson. He stood on the mosaic and delivered a speech, part of which read:

> My painting for the mosaic in the forecourt of Parliament House represents all the indigenous people in this land, the wider Australia. That's why I put all the different animals – represent to me all the peoples at this place . . .
>
> This has all been changed. This is no longer a meeting place for Aboriginal people. The government of Australia are still not recognizing our people and our culture. It is abusing my painting and insulting my people. It makes my people sad that government does not respect my painting or my people. I want to take my painting back for my people (quoted in Johnson 1997: 127).

At the conclusion of his speech, Nelson put down the megaphone and took up a hammer and chisel. Protest organizers had planned for Nelson to dig out the centre stone of his mosaic. In the event, he did not do this, but the symbolism of him standing above the stone with a hammer and chisel was enough to guarantee the protest significant media and political impact (see Figure 3.3: similar photographs appeared on page one of *The Sydney Morning Herald*, *The Age* and *The Australian* on 28 September 1993).

In both the 1988 and 1993 mosaic protests, the inclusion of a significant work of Aboriginal art in the publicly accessible space around Parliament House provided Aboriginal people with an opportunity to

Figure 3.3   Michael Jagamara Nelson threatens to chisel out the centre stone of the Forecourt Mosaic during Native Title protest, 1993 (source: *Canberra Times*; used with permission)

contest their treatment by the Commonwealth Government. In challenging the official meaning of the mosaic, they challenged the very identity of the space in which it is located. But in doing so, they have raised complex questions about who has the authority to use the mosaic as a space for representation. Gilbert's use of the mosaic was challenged both by politicians, who immediately declared the mosaic off-limits to future protestors, and by the artist himself. In 1993, Nelson used his own authority as artist, and the Presiding Officers and APS did not intervene to prevent his protest going ahead. But even his status as artist was not unchallenged. The same women who had dedicated the mosaic to its Dreaming purpose in 1988 were there at the 1993 protest, urging Nelson not to deface the mosaic and commit 'a crime against the laws of western desert society' (Johnson 1997: 20).

They came and they stayed: alternative embassies outside Parliament House

A number of protestors have come to Parliament House and stayed, sometimes for a period of several days, other times for even longer periods. The
Aboriginal Tent Embassy first established outside the provisional
Parliament House in 1972 has been a model of protest taken up by a variety
of movements. Protest structures have presented a range of dilemmas for
the Presiding Officers and for the NCA, who are responsible for managing
the area of Federation Mall which faces the new Parliament House but is
beyond the Presiding Officers' jurisdiction.[3]

An environmental alliance involving the Greens and the Wilderness
Society organized a 'Forest Embassy' on Federation Mall from 4–8
November 1994 (see *Sydney Morning Herald*, 8 November 1994: 8).
According to Lucy Horodny of the Greens, the Forest Embassy served two
purposes:

> first, obviously, to apply political pressure on the Federal Government, and,
> secondly, to provide a public education campaign. In other words, the event
> is staged to put direct pressure on the politicians in this House, which is why
> the staging of the event is held in front of Parliament House. But it is also
> staged to entice local Canberrans to be involved in the forest embassy and
> to educate people about forest issues (Joint Standing Committee on the
> National Capital and External Territories 1995b: 506).

The organizers of the Forest Embassy went to significant lengths to
gain permission from the NCA for the establishment of structures on
Federation Mall for a period of three nights. Negotiations with the NCA
for a permit were protracted and difficult, but some structures were
eventually authorized. No authorization was given for any protestors to
stay overnight on Federation Mall. To make any such camping difficult
and uncomfortable, public toilet facilities in the car park underneath
Federation Mall were locked at night during the course of the Forest
Embassy protest. The NCA's reasons for making a protest of this duration so difficult did not impress the protestors. As one organizer from the
Greens noted:

> Particularly what struck the people who were protesting at the forest embassy
> was the great concern expressed by the authorities about the possibilities of
> damage to the grass. It was extremely ironic that vast tracts of old growth
> forest were being clear-felled and huge numbers of endangered species were
> being wiped out with no apparent concern from the authorities, whereas two
> species of imported grass were being treated with such reverence (Joint

Standing Committee on the National Capital and External Territories 1995b: 381).

The conduct of medium- or long-term protests has been particularly contentious at Parliament House. The Green coalition in this particular protest attempted to 'do the right thing' by negotiating an agreement with the NCA. Others, however, have simply arrived and stayed, with mixed results.

Three days after the Forest Embassy was packed up, Joseph Bryant and his supporters arrived at Parliament House with the intention of staying until the Commonwealth Government had listened to their grievances. Bryant wanted increased regulation of the banking system to protect farmers and rural families in hardship. A large metallic 'Trojan Horse' was towed to Canberra on a trailer, and the Trojan Horse and basic camping facilities were set up in Federation Mall. Unlike the environmentalists, Bryant had not sought permission for these structures from the NCA. The Trojan Horse stayed at Parliament House for over five months. Bryant's reasons for staying so long were simple: 'no interest in the issues or the reasons for us being there was shown, so we stayed' (Joint Standing Committee on the National Capital and External Territories 1995b: 465).

Eventually, after a request from the Presiding Officers, the NCA instructed the AFP to remove Bryant and his structures from the Mall. There had been some reluctance to act until this point, partly because of the apparent inconsistency of removing Bryant while leaving the Aboriginal Tent Embassy outside Old Parliament House undisturbed. The Aboriginal Tent Embassy had been re-established by Aboriginal protestors on its original site in 1992, on the twentieth anniversary of the first Aboriginal Tent Embassy. Since 1992 the Aboriginal Tent Embassy had expanded to include sheds, a post box, a mural and a growing array of ceremonial structures (*Koori Mail*, 12 February 1992: 1, 5; 3 June 1992: 27 (Australian Heritage Commission 1995). In their letter asking for Bryant's eviction in 1995, the Presiding Officers attempted to help the NCA resolve this apparent inconsistency:

> We are aware that the Aboriginal 'Tent Embassy' is similarly illegal and that to remove the structures in front of the Parliament House without addressing the issue of the 'Embassy' would lay the authority open to accusations of double standards. However, if the issue of the 'tent embassy' is not to be confronted we believe that a distinction can be made between the ceremonial Mall and less sensitive areas of the Parliamentary Zone . . . [T]he structures in the Mall provide a continuing and inappropriate backdrop to state and parliamentary occasions whereas the 'Tent Embassy' is relatively unobtrusive

(Joint Standing Committee on the National Capital and External Territories 1995b: 199).

Bryant and his Trojan Horse were removed at 10 p.m. on the evening of 27 April 1995 by the AFP. It was not only the duration of Bryant's protest that led to action. Authorities were also concerned that others would be encouraged to conduct similar protests if Bryant's protest appeared to be sanctioned through inaction. Such fears were exacerbated by Barry Williams of the Lone Fathers' Association, who had established the 'Embassy for the Australian Family' on Federation Mall early in 1995. This embassy was also removed by police on 27 April. Williams later accused the Presiding Officers and the NCA of hypocrisy:

> It is now patently obvious that if you are of a certain race in this country you have the right to demonstrate in the manner you consider to be the most effective. The Government has demonstrably shown that other Australians do not have the same rights (quoted in Joint Standing Committee on the National Capital and External Territories 1995b: 191).

Inquiry into the Right to Protest begins, 1994

During the controversy over these protests, the 'Inquiry into the Right to Legitimately Protest on National Land, and in the Parliamentary Zone in particular' (its full and rather inelegant title – hereafter 'the Inquiry') was announced by the Labor Commonwealth Government. The 'ongoing protest represented by the Aboriginal Tent Embassy' played a significant part in the Government's decision late in 1994 to launch the Inquiry (Howe 1994). Deputy Prime Minister Brian Howe acknowledged that 'the Tent Embassy was significant to many Aboriginal people and the issues surrounding its continued existence are both complex and sensitive' (Howe 1994). For this reason, he argued that there was 'a need to formalize [the right to protest or demonstrate], to clarify the legal position and scope for these types of demonstrations and to adopt guidelines to assist those administering the area' (Howe 1994). In announcing the Inquiry, Howe also made reference to 'environmentalists and farmers demonstrating in the area and some illegally camped in the front of Parliament House'. There was a growing sense of concern that unless the Commonwealth Government could establish a clear framework for the management of protest, the example set by the Aboriginal Tent Embassy at Old Parliament House stood to be replicated in front of the new Parliament House. Finding a way to prevent this, while simultaneously recognizing the status of the Aboriginal Tent Embassy, became a significant issue for the Commonwealth Government.

They came and they stayed (with trucks): the loggers' blockade, February 1995

By the time a call had been made for public submissions to the Inquiry, another memorable protest had occurred at Parliament House. On 29 January 1995, more than a hundred logging trucks surrounded Parliament House, lining the feeder ramps and Parliament Drive. The drivers, along with timber workers and their families from the South Coast of NSW, blockaded Parliament House for one week. They were in Canberra to protest against Commonwealth policies on the granting of woodchip licences, which they argued would reduce employment in the timber industry. Their action, according to the *Sydney Morning Herald* (2 February 1995: 1), 'proved that Canberra's $1.1 billion Parliament House has been a siege waiting to happen ever since it was built'.

The blockade prevented vehicular access to Parliament House to all but emergency vehicles, and forced visitors, Parliamentarians and Parliamentary staff to enter the building on foot. Protestors and their families camped on the lawns, children played in the forecourt fountain. Some who were lucky enough to be signed in as visitors by sympathetic Members of Parliament took showers in Parliament House itself, upsetting the designed separation of the public from Parliamentarians. The blockade clearly contravened the existing restrictions on protest at Parliament House, as well as numerous traffic regulations, and the organizers had not sought permission for the protest. However, no attempt was made by the APS or the AFP to enforce these laws and regulations by forcibly removing the trucks. No doubt this was largely for practical reasons – as one journalist remarked:

> Short of calling in a battalion of tanks, there appears not a thing the nation's legislators, the police, or anyone else can do about it (*Sydney Morning Herald*, 2 February 1995: 1).

Blockade organizers instead liaised daily with the Presiding Officers and AFP.

The blockade was broken up voluntarily after a week, when Prime Minister Keating amended his proposed woodchip policy. The breach of restrictions which limited protest to areas outside Parliament Drive, with trucks, tents and barbeques overwhelming the public spaces of Parliament House for several days, was fundamental to the success of the protest. One of the protest organizers later remarked when giving evidence to the Inquiry that:

If we had just been there a day or so, waved a few placards around, gone away, I think there would be a lot more unemployed forest workers and a lot more seriously affected communities out there (Joint Standing Committee on the National Capital and External Territories 1995b).

The protest very effectively challenged the authorized identity of the space around Parliament House – it disrupted both the symbolism of Parliament House and the everyday functions of Parliament. The sight of public servants, politicians and international dignitaries having to walk past the blockade every morning on their way to work featured nightly on the television news. The blockade also significantly disrupted deliveries of basic materials like paper and food which were part of the efficient functioning of the building.

The response of the police and Presiding Officers to the loggers' blockade led to more accusations of double standards in the regulation of Parliament House, this time from those involved in the Forest Embassy. A representative of the Wilderness Society later told the Joint Standing Committee conducting the Inquiry:

we went through this lengthy process to get this piece of paper that gave us the permission to be out here in front of this place for five days. I believe that we abided by the law. We went to a lot of trouble to try to ensure that we did not disrupt this place in all sorts of ways, for example, by making sure that speakers were not facing towards Parliament House and all those sorts of things. Then at the end of January there was the loggers' protest, which was highly disruptive, not regulated and not policed . . . It made me feel that there were no regulations being imposed on that protest, yet we had to jump through a lot of hoops to stay (Joint Standing Committee on the National Capital and External Territories 1995b: 508).

After the event, the Presiding Officers responded by establishing the formal *Guidelines* which restricted protest to the designated area at the front of the building, and banned any protests which would encircle or obstruct entry to the Parliament. This response seems ill-considered – if they or the AFP had the will, they could have legally ordered and enforced the removal of the trucks under existing traffic regulations. It was not as if the blockade had cleverly exploited a loophole in the existing *Guidelines*. Rather, it had been a sizeable and effective show of force, using large trucks to make a point by tactically exploiting a vulnerability in the design of the building. Nonetheless, the amended *Guidelines* further restricted the spaces in which protest could be conducted by other protestors who did not arrive with a large fleet of logging trucks. They also included measures designed to address problems caused by other, less spectacular protests. For example, the stipulation that speakers could not face the Parliament was designed to

prevent a repetition of protests which had noisily interrupted official cere-
monies in the forecourt.

The Cavalcade to Canberra, or, 'the Parliament House Riot', August 1996

Just as the Joint Standing Committee was finalizing its deliberations regarding the management and regulation of protest, one of the largest protests
to occur at Parliament House would bring a new urgency to the Inquiry.
The Coalition Commonwealth Government elected in March 1996 had
quickly provoked the ire of union and community sector groups with its
programme of industrial relations 'reform' and cuts to public sector spending. The Australian Council of Trade Unions (ACTU) organized the
Cavalcade to Canberra on 19 August 1996 with the intention of putting
public pressure on the Government, by bringing unionists together with
community and indigenous groups in a show of general community opposition to the Government's direction. It was hoped that up to 30,000 people
would attend, and buses and trains were organized for participants from
Melbourne, Sydney, Wollongong and other cities and regions.

According to some accounts, the target of 30,000 people was reached.
ACTU Secretary Bill Kelty declared that it was 'the most successful rally
in the history of the nation' (*The Age*, 20 August 1996: 12). But it wasn't.
While Kelty may have been pleased with the turnout, his comments served
only to provoke the mass media. They were much more interested in the
scuffles involving a group of protestors who had broken away from the main
rally and attempted to force their way past police guarding the entrance to
Parliament House. *The Australian* was not alone in dubbing these scuffles
a 'riot', and asked:

> Do the rioters reject the democratic process? Their violence against the prin-
> cipal democratic institution in this country – violence that saw blood spilled
> on the marble floor and walls of Parliament House – certainly shows a fright-
> ening disrespect for the national symbol of democracy and a disregard for the
> result of free and fair elections (*The Australian*, 20 August 1996: 14).

The ACTU's plans for the rally had fit neatly within the new *Guidelines*
regulating protest activity. As negotiated with the Parliament's Security
Controller and the AFP, a stage had been erected in the designated protest
area at the top of Federation Mall, facing down the Mall away from
Parliament House. However, while thousands listened to speeches from a
range of community, union and political leaders, others tried to force their
way through the front doors of Parliament House. Some protestors and

Figure 3.4   Protestors hoist their own flags on Coat of Arms at Parliament House during the Cavalcade to Canberra protest, 1996 (photo: Kurt Iveson)

police officers were injured in the altercations, and damage was done to the doors and the Parliament House Gift Shop. While most police were distracted with these scuffles, many other protestors committed more minor infractions of the *Guidelines*. They peacefully filled the forecourt, walked on the grass over the top of the Parliament, and some managed to climb up to the Coat of Arms above the public entrance to hang banners and signs (see Figure 3.4). Many people listening to the speeches were unaware of the fracas until after the event, while others were made aware of it when one of those who had attempted to enter Parliament House leapt to the official podium showing off a police shield that had been 'liberated' in the scuffles.

The media coverage of the protest focused almost exclusively on the violence, and both the media and the Government sought to attribute blame to the ACTU as rally organizer. Newspaper headlines the next day such as 'Rioters storm Parliament' (*The Australian*, 20 August 1996: 1), 'Parliament besieged' (*The Age*, 20 August 1996: 1) and 'Bloody Protest' (*Canberra Times*, 20 August 1996: 1) were all accompanied by graphic pictures of the scuffles and bloody-faced protestors. Editorials proclaimed 'Canberra riot a disgrace' (*The Australian*, 20 August 1996: 14) and labelled it 'The ACTU's Responsibility' (*Sydney Morning Herald*, 20 August 1996: 13). After inspecting the damage to the Gift Shop, Prime Minister John Howard cut short a planned meeting with ACTU President Jenny George and said:

never under any circumstances will my Government buckle to threats of physical violence or behaviour of that kind (quoted in *Canberra Times*, 20 August 1996: 1).

The *Sydney Morning Herald* rammed home Howard's assessment, and had the following suggestion for the union movement:

The ACTU will lose – on the Workplace Relations Bill and any other issue – if it abandons reasoned argument for the blunt and dangerous weapon of mass demonstrations (*Sydney Morning Herald*, 20 August 1996: 13).

Some weeks after the rally, the ACTU issued a public statement expressing regret at what had happened, which concluded:

While the actions of the tiny minority have undoubtedly done harm to the collective union movement, the extent of the union and community opposition displayed that day highlights our determination to continue to campaign in opposition to the Howard Government's industrial relations legislation (quoted in Norington 1998: 302).

But after weeks of mainstream attacks on the union movement, these statements and others like them in union and community journals had little effect in changing the 'truth' of what had happened, even within the union movement itself, let alone in the wider community. The rally was to have been a show of numbers for the media to back up the position of ACTU leaders in their meeting with the Prime Minister later in the day. But while union leaders had liaised closely with the Parliamentary authorities and police before the rally on behalf of the protestors, they clearly did not have control over the diverse range of union and community activists who came to Parliament House for the protest (Iveson 2001). The actions of those who broke away from the organized rally gave the media a chance to focus on their violence instead of the claims of the ACTU leaders.

Finding the Right 'Balance': The Inquiry into the Right to Protest

Debating the 'special qualities' of Parliament House and the National Capital

If these protests (and others) can be considered as skirmishes between Parliamentary authorities and protestors over the use of the Parliamentary precinct as a venue for protest actions, then the Inquiry provided an opportunity for these groups to articulate more carefully claims about the place of protest at Parliament House. Here, we can see quite clearly the interaction of the three urban dimensions of public address discussed in Chapter 2. The Inquiry was part of the ongoing project of investing Parliament House with a spatial identity. It specifically sought to make the possibilities of this particular place for different forms of public address the object of public debate. Further, the Ministerial call for submissions made it clear that the terms of this public debate were to be framed with reference to a vision of Canberra as a particular kind of city/public – Canberra was positioned in relation to the collective subject of 'the nation' as 'the National Capital':

> Given the right to protest or demonstrate is considered to be a legitimate activity, the Minister has asked the Committee to examine, in detail, how freedom of expression may be allowed without compromising the special qualities of the National Capital (Joint Standing Committee on the National Capital and External Territories 1995a).

This reference to the 'special qualities' of the National Capital reflected a growing concern that protests had the potential to challenge the identity of the spaces that had been so carefully designed and established by Giurgola and the PHCA.[4] The Inquiry into the staging of protest, then, was from its inception also an inquiry about the nature of the identity of Parliament House and its place among the 'special qualities' of Canberra, the National Capital.

Representatives of a variety of state agencies expanded on the special qualities of Parliament House in their written and verbal submissions to the Inquiry, and attempted to reassert and clarify their authority to determine the identity of the space in question. Almost universally, they expressed the desire to find a 'balance' between the rights of protestors and the various needs of the Parliament and Executive. The Office of Prime Minister and Cabinet expressed their desire to ensure that ceremonial occasions were free to occur in the forecourt without disruption. The Department of Foreign Affairs and Trade wanted to ensure that the 'dignity' of foreign politicians and officials was not violated by protest, and justified their position with reference to obligations under the Vienna Convention. The Presiding Officers were anxious that access to Parliament for politicians, staff and visitors not be obstructed by protestors (Joint Standing Committee on the National Capital and External Territories 1995a). The Attorney-General articulated the need for more clarity in the legal framework for the management of protest:

> Law enforcement agencies have encountered difficulties in identifying the powers available to them to regulate protests or demonstrations in the zone (Joint Standing Committee on the National Capital and External Territories 1995a: 353).

These law enforcement agencies – the AFP and the APS – made their own submissions which sought clarification of their powers and provided advice about the practicalities of different regulatory and management options. Both of these agencies suggested that a permit system for protests be introduced, to ensure that protestors liaised with the Parliamentary Security Controller. It was argued that this would assist the AFP and APS to gather intelligence and operationally plan for protests, by forcing protestors to make commitments about the conduct of protest in advance. They also argued that the maintenance of an agreed protest space made the regulation of protest much easier. The AFP noted that:

> Rightly or wrongly, we have set a standard for the demonstration area at the front of Parliament House. We are trying to avoid the problems around the building. We just do not have the resources to be at every point of the building, so it is better to try to keep it so (Joint Standing Committee on the National Capital and External Territories 1995b: 142).

The NCA added their voice to calls for clarification of laws regulating protest, in particular in relation to their role as managers of national land. Like good town planners, they hoped to influence protestors' behaviour by redesigning the available protest space:

> In the case of a standard protest kit, if you like – perhaps a couple of marquees that could be a first-aid tent and an information centre, and a stage that either comes up out of the ground or is stored below in the car park and brought out on request when a demonstration is to take place – that would be one means of facilitating demonstrations and focusing people's attention on physical structures that would in turn affect social behaviour (Joint Standing Committee on the National Capital and External Territories 1995b).

There were no submissions about the need to protect the Parliamentary grass.

While the various government and law enforcement agencies did not speak with one voice, the attempt to balance the right to protest with the needs of Parliament House produced a picture of 'legitimate' protest with a quite specific time–space axis. A protest should be of a short duration. It should not interrupt ceremonial occasions, nor should it inconvenience Parliamentarians, staff or visitors in any way. Preferably, it should be well organized, and protest plans ought to be negotiated in advance with the APS or the AFP. A protest should, of course, be orderly. And a protest ought to make its point symbolically, through the media, rather than directly to visitors or Parliamentarians, who should have not have to face protestors at the various entrances to Parliament House. In essence, public protest was not to disrupt the other public functions of the building. While there was general agreement on the dimensions of legitimate protest, whether such protest could best be assured via a permit system remained a matter of debate.

The dimensions of legitimate protest proposed by state agencies were challenged by protest organizers, academics and various non-government agencies who gave evidence to the Inquiry. There was near-universal condemnation of any suggestion that the 'special qualities' of Parliament House could justify restrictions on protests. The Queensland Law Society, for example, argued that:

> whatever the 'special qualities' of the Nation's Capital may be, they be recognized only to the extent that can be allowed without compromising freedom of expression (Joint Standing Committee on the National Capital and External Territories 1995a: 150).

Other lawyers, such as Elizabeth Evatt and the Human Rights and Equal Opportunities Commission, made reference to the *International Covenant on Civil and Political Rights*, and argued that clear rights to protest should be enshrined in legislation, rather than simply acknowledged in practice, in order to ensure that administrative convenience was not allowed to outweigh rights to speech and assembly. Activist and lawyer Tim Anderson argued against the notion that rights to protest had to be 'balanced' with other considerations:

> This is to posit a fundamental right as simply another element to be weighed up, given out, or withdrawn as part of some administrative balancing act . . .
> It is precisely because governments repeatedly seek to tamper with fundamental rights, for reasons of their own administrative convenience, that demands were made for such rights to be acknowledged internationally, as matters beyond the grasp of governmental convenience (Joint Standing Committee on the National Capital and External Territories 1995a: 302).

The Greens and Joseph Bryant expressed their frustration at being restricted to a designated area that was some distance from the entrances to the building used by politicians. Rather than focusing on the inconsistency of the existing law, as did state agencies, they and other protestor groups argued that the existing laws were adequate, and problems were the result of the inconsistent *application* of those laws.

In their final report, the Joint Standing Committee accepted the argument that no extra burdens should be placed on those wishing to stage protests at Parliament House. They recommended against a permit system, arguing that 'a permit system may introduce new complexities without necessarily resolving existing difficulties' (Joint Standing Committee on the

National Capital and External Territories 1997: 70). However, they did support 'the introduction of a [permit] system for the management of protest structures erected on national land'. Not surprisingly, the Committee also agreed with most state agencies who argued that the existing arrangements did not place undue burdens on protestors. The notion of achieving 'balance' informs the recommendations in their final report:

> The challenge is to ensure that the legislative and administrative arrangements governing the conduct of protests on national land achieve an appropriate balance between the right to undertake protest activity and the obligation to ensure that any such activity does not endanger public order, public safety or the rights of other users of national land (Joint Standing Committee on the National Capital and External Territories 1997: xvi).

There were no recommendations to amend the existing *Guidelines* of the Presiding Officers. The Committee preferred to 'build on existing cooperative arrangements for dealing with protests' (1997: 70) which were usually established through discussions between protestors and security agencies prior to the conduct of protests. One of the main recommendations to this end was to produce a public information booklet, which would:

> encourage prior notification of protests to a protest coordination officer of the Australian Federal Police and which would provide relevant information and advice on the conduct of protests in the national capital, and focusing on national land (Joint Standing Committee on the National Capital and External Territories 1997: 72).

This recommendation was adopted and a booklet was eventually published by the NCA in June 2000.

The Form of 'Legitimate Protest'

The findings of the Committee, and the *Guidelines* subsequently refined and published by the Presiding Officers and the NCA, clearly demonstrate that the priorities of Parliamentary authorities were accorded the highest priority during the Inquiry. Public input was welcomed to the extent that it helped to clarify how protest could best be managed, but defining and protecting the identity and 'special qualities' of the Parliament remained the preserve of the agencies responsible for its management. The model of 'legitimate protest' that has been articulated through the final report and is enforced with reference to the *Guidelines* serves to privilege a very particular form of public address by protestors – one which is premised on a narrow, normalizing understanding of how public address and politics more generally are to be conducted. Two key aspects of this model are worth considering in further detail.

First, expectations about the kinds of protest that are 'in place' at Parliament House are underpinned by an assumption that the staging of protest is purely a matter of collectively representing pre-formed claims, rather than a matter of collective claim formation and deliberation. The formation of claims and values is assumed to take place somewhere else, prior to their representation through the staging of protests. By allowing protestors the time and space to be *seen*, the various land management and regulatory authorities felt that they had met their obligations concerning the right to protest. These expectations are reinforced by the *Guidelines* in several ways. They make protest structures difficult to establish. This frustrates protestors who wish to use the space around Parliament House to develop or *form* their opinions by establishing any physical structures that will sustain interaction among members of a protest group over a lengthy period. An NCA official, when asked what he thought was a reasonable time-limit for protest structures, argued that 'somewhere in the week to three week vicinity is a pretty reasonable time. Any protest that extends beyond that period probably loses its impact,' he believed, because 'all members of Parliament will have seen it in the space of a week or three weeks', and 'the media have probably lost interest after the first few days' (Joint Standing Committee on the National Capital and External Territories 1995b: 21). The notion that claims and values should be established 'somewhere else' prior to protest events at Parliament House is also reflected in the *Guidelines'* expectation that protestors are well organized, with an established hierarchy of leaders who are in a position to negotiate with Parliamentary authorities and police in advance of the staging of a protest. Preferably, leaders should be able to make and enforce commitments on behalf of protestors. This was the main criticism levelled at the ACTU by police and others, following the Cavalcade to Canberra.

Second, the time–space axis of legitimate protest also rests on the liberal assumption that if protestors are given an opportunity to *speak*, this will be enough for them to be *heard* if they have a legitimate point to make. Stephen Martin, one of the Presiding Officers, told the Inquiry that:

> What we have tried to say is that, by having a demonstration area where people can get their point of view across, generally speaking, that [balance between right to protest and other administrative needs] is achievable. Most people who come here, if they have a legitimate beef,[5] arouse the interest of the media. After all, that is what they are looking to do. They want to capture that ten second grab on the nightly news so that they can get their point across (Joint Standing Committee on the National Capital and External Territories 1995b: 195).

Protests at Parliament House which attempt to put claims directly to other members of the public entering the building, or to politicians (all of whom use entrances where protest is forbidden), are prohibited. The use of the space around the Parliament by protestors to put their point to politicians must remain 'symbolic', not direct. It is as if the public sphere is a kind of 'free market' for ideas and opinions which stand or fall solely on their merits, much like commodities in the marketplace (Mitchell 1996). The identity or the status of participants does not matter, only the strength or rationality of their ideas. As a consequence, if protestors have a 'legitimate beef', simply gathering in the designated protest area should be enough to attract the attention of the media and through them the wider public sphere. From this perspective, the restriction of protest to the designated area is not an undue burden on protestors, as long as it is applied equally to all protestors.

After the *Guidelines*

The model of 'legitimate protest' defined by the *Guidelines* and reinforced through the Inquiry is part of the identity of Parliament House. This has significant consequences for those who wish to use the public space around Parliament House for the conduct of protest. Any protest which contravenes the *Guidelines* – because, say, it is conducted over several weeks, or it uses spaces outside the designated protest area, or it has no 'leaders' who are in a position to make commitments on behalf of protestors – will result in a confrontation with land managers and regulators. Such a skirmish will often take attention away from the content of the claim, and may even be used to undermine its validity. Protestors are inevitably characterized as attempting to 'force' an idea into the public sphere, and from the dominant liberal perspective, if ideas 'must be forced into the marketplace, they are by definition invalid' (Mitchell 1996: 164). But confrontations are inevitable if protestors seek to tactically use the opportunities for protest provided by Parliament House in ways that facilitate alternative, counter-public modes of formation and representation. Managing these confrontations thus becomes a key tactical concern for protestors.

A range of protests have successfully worked to construct forms of public address which are consistent with the *Guidelines*. In particular, the designated protest space has been used quite effectively by some activists who have sought the kind of media coverage envisaged and endorsed by the existing regulatory framework. Here, public address is literally staged with subsequent mediatized circulation in mind. Several protests have used Parliament House as an effective visual backdrop for protest actions designed to circulate claims via the mass media. For example, advocates of 'reconciliation' with Australia's indigenous peoples installed a 'Sea of

Figure 3.5   Sea of Hands protest Federation Mall, 1997 (photo: Kurt Iveson)

Hands' in Federation Mall in 1997, with each plastic hand planted in the ground containing the signature of one person who had signed a petition in favour of an official apology from the Commonwealth Government to the 'stolen generation' of Aboriginal people taken from their parents by state authorities (see Figure 3.5).[6] When the status of the Aboriginal Tent Embassy outside Old Parliament House was questioned by the Minister for Aboriginal Affairs in 1999, its occupants staged a protest in the designated protest area involving the lighting of a ceremonial fire, which drew considerable media attention to their plight (see Figure 3.6). In 2003, activists critical of Australia's participation in the invasion of Iraq installed a mass of stuffed body bags to make an anti-war protest.

Nonetheless, the consolidation of the *Guidelines* for protest has by no means settled the question of how Parliament House is put to use as a venue for staging protest. Protestors continue to explore the opportunities that Parliament House affords, testing the limits which Parliamentary authorities have attempted to establish and enforce through the *Guidelines*. In particular, some activists have realized that so long as the building and

Figure 3.6   Tent Embassy protest with media photographers in Federation Mall, 1999 (photo: Kurt Iveson)

its precincts remain open to 'visitors', 'visiting' provides its own possibilities. One can stage a protest on the 'inside', putting a case directly to Parliamentarians and visitors – not by battering one's way through the front doors against phalanxes of police, but rather by simply refusing to identify as a 'protestor' until one enters the building. And so, politicians have been heckled from the public galleries by individuals and groups, all of whom are duly escorted from the building by Australian Protective Service officers as soon as they begin to make a noise or unfurl their banners. While many hecklers appear to be acting as individuals, sometimes exasperated and sometimes inebriated, others have clearly been collectively organized. During the debates over amendments to Native Title legislation following the High Court's Wik decision, some 'visitors' in the public galleries of Parliament stood up, removed outer layers of clothing to reveal t-shirts with pro-land rights slogans and began to shout their disapproval of the Howard Government's approach. As each one was ejected, another stood up to take their place. In August 2002, four Greenpeace activists made their way to the publicly accessible rooftop of the Parliament, and then began to scale the 80-metre flagpole – draping it with a banner urging the Commonwealth Government to ratify the Kyoto protocol on greenhouse gas emissions (*Canberra Times*, 21 August 2002: 3).

In another such event on 18 March 2003, as Australian Prime Minister John Howard rose to his feet in the Parliament to announce that Australian troops would participate in the invasion of Iraq, the voices of around 150 women rose in song to fill the entrance hall of the Parliament with a lament. As one of the two women who wrote the lament and initiated this action later reflected:

> we scattered ourselves over the huge foyer of the Parliament so we didn't look like a choir. None of us knew everyone or how many we were (Cloughley 2003).

Such protest tactics bear more than a passing resemblance to other protest efforts that have attracted critical interest in recent years, such as the Reclaim the Streets street parties/protests that have taken place first in London and subsequently in other cities around the globe since the early 1990s (Jordan 1998). Like the protestors/partygoers who materialized on the middle of the M41 in London as if from nowhere only to disappear several hours later, the Chorus of Women also materialized as a collective 'protest' in the moment of the action itself. This is not to say that such actions are random or unplanned, but rather that participants have figured themselves rather differently in relation to static conceptions of 'protest' such as those embodied in the *Guidelines*. They refuse the identity of the 'protestor' who must announce their intentions by sending deputations to Parliamentary authorities. They reject the view that protest must at all times be 'rational' in the sense that it involves the making of claims in ways that fit with established norms of political dialogue and debate.

Conclusion: Authorizing Protest at Parliament House

Protests at Parliament House take their place alongside a variety of other forms of public address. Different publics are imagined and addressed through speeches to Parliament, Parliamentary press briefings, the artworks and architecture of the building, lunchtime lectures in the auditorium, to name but a few. Thus, the designation of the Parliament as a 'public building' and the identification of some of its spaces as 'open to the public' mobilize quite different (and in some ways contradictory) meanings of 'publicness'. Like the other forms of public address which are sustained through the site, political protests have been assigned a place in the Parliamentary precinct. As I have shown in this chapter, however, the specifics of the place assigned to protest have been a constant problem for the designers and managers of the precinct and for protestors alike. We can quite clearly observe the interaction of the three urban dimensions of public address identified in Chapter 2. The use of the Parliamentary precinct as a venue for protest has become the object of governmental concern and public debate, in a Parliamentary Committee of Inquiry and in the mass media more generally. And such considerations have taken place with reference to wider concerns about the city of Canberra as the National Capital, with whatever 'special qualities' that implies. The designers and regulators of Parliament House view protest as a legitimate function that fits with the identity of the building as the centre of the National Capital and the home of Parliamentary democracy. But this identity has also given them cause to define the form which 'legitimate' protest should take quite narrowly.

Parliamentary authorities have attempted to make protests governable by allocating them a time and a place within the Parliamentary precinct, articulating and policing a set of rules which curtail the forms that protest can take in order to 'balance' contestatory forms of public address with other public functions and uses of the precinct. From their perspective, the *Guidelines* do not compromise the identity of Parliament House as the symbolic and functional home of Australian representative democracy because they give all Australians an equal opportunity to stage a protest there and do not place undue restrictions on the content of claims that can be made. This is achieved by the provision of an authorized protest space and the consistent application of the rules to all protestors. As we have seen, some protestors have attacked the regulatory framework as unjust on the grounds that these rules have in fact been applied inconsistently. This ranges from complaints about the unequal treatment of Aboriginal protestors to concerns about the unequal treatment of loggers. In responding to these claims, the various authorities responsible for land management and regulation have sought to clarify the rules in order to better facilitate their consistent application to all protestors.

But, even in their consistent application, the *Guidelines* work to frame different forms of public address hierarchically relative to one another. Protest is positioned in a subordinate relationship to other forms of public address that work through this site such as Parliamentary debate, with the concept of 'Parliamentary privilege' enshrined in law. But the regulation of protest is not only a matter of positioning protest relative to other public functions. It has also been a matter of privileging some forms of protest over other forms. The rules for protest claim an impartiality in the *universality* of their application, but in fact they embody a *particular* form of protestatory public address that is not necessarily available, accessible and/or relevant to all who would seek to make a claim at Parliament House. Rather than accepting that 'protest' may involve a potentially wide range of forms, the *Guidelines* limit 'legitimate protest' to a very specific time–space axis. Some protestors – particularly those who wish to stage protests which make mediated claims on behalf of a stable group – are privileged over those protestors who may wish to put their claims to people (Parliamentarians, visitors, etc.) directly, or who want to use public space as a space of public formation as well as a space of representation. The hierarchical framing of protest forms relative to one another truncates the exploration of heterogeneous forms of public address at this important site.

To put this another way, Parliamentary authorities seek to contain the social and spatial imaginaries of the various publics who may seek to make a claim through the spaces of Parliament House. Of course, as I have shown, they are only partially successful.

From the critical perspective on protest at Parliament House I have developed in this chapter, we can see that efforts to restrict or widen opportunities for the making of claims through this place are only in part a simple matter of access. Some parts of the Parliamentary precinct are indeed 'off limits' to protestors, and this continues to be a source of frustration for many activists. But this limitation is said to be addressed through the provision of a designated space that is 'open to all'. With respect to this situation, the question of *authority* assumes a great significance. By what authority are the terms of access established and policed? The Presiding Officers and the NCA have quite literally sought to establish themselves as the sole authors of Parliament's identity, relying on legislative powers granted to them under Commonwealth law. While for a time the rules for protest were made the object of public debate through a Parliamentary Inquiry, this Inquiry never put the structures of authority into question – rather, it opened up a limited range of questions about how existing authority might be exercised. Nonetheless, some protestors have found ways to contest the rules by claiming other sources of authority with respect to the uses of Parliament House and its precincts. In 1993, for instance, Michael Jagamara Nelson claimed his authority as the Forecourt mosaic's artist to use it as a site of protest, while loggers created their own authority with the use of heavy machinery. Perhaps more widely available as an alternative source of authority is the very identity of Parliament House as a building which is supposed to serve as the symbolic and functional property of 'the people/nation' that is represented by the Parliament. The massive investment in the identity of Parliament House has been redirected by those activists who have claimed an authority to act as citizens of 'the nation' that the building is supposed to symbolize and functionalize. By staging protest actions at Parliament House in the spaces zoned for 'public use', they have either ignored or resisted the notion that they must seek legitimation from Parliamentary authorities to stage a protest. In so doing, they are effectively seeking to establish a new understanding of the relationship between of 'the public' that is territorialized in the 'public entrance hall' and the 'public galleries', and 'the public' that is symbolized by the architecture and identity of the building. The public in the gallery and the entrance hall is no longer a passive witness to Parliamentary debate, but an active participant in political dialogue about the nation's future. Until, of course, they are ejected and put back in their proper place as soon as they are identified as a 'protestor' (rather than a 'visitor') by Parliamentary security agents.

Of course, some governance of protest at Parliament House is inevitable given other demands on the place. But concerns with the 'balance' of uses might be pursued more democratically through the establishment of permanent institutionalized dialogue about protest forms premised on the dynamic heterogeneity of public address. Rather than presuming to settle the question of what kind of governance is appropriate in advance, such dialogue might seek instead to allow the possible uses of the Parliamentary precincts to be debated from a range of perspectives. Rather than securing the authority of some agents to decide what is appropriate through legislative means, such dialogue might instead open up the question of authority itself to public debate. This institutionalized public dialogue would not be without its own difficulties and limits. In particular, any attempt to establish a more democratic governance structure would reveal the extent to which the question of 'authority' is not only a matter of dispute between protestors and land management agencies – it is also a matter of dispute *within* protest groups. Who would be 'authorized' to represent different publics, in negotiations between protest organizers and governance agencies? As the ACTU leadership found out to its cost, any claims to representational authority within protest movements can be confronted and undone by the actions of protest participants who refuse to conform to leadership expectations.

Canberra's imagined identity as the National Capital is both an impediment and a resource to the democratization of protest opportunities at Parliament House. The 'special qualities' of Canberra as the National Capital have been invoked by regulatory authorities in their efforts to establish and justify the existing *Guidelines*. And yet their efforts to ascribe an identity to the new Parliament House as the symbolic and functional 'centre' of Australian Parliamentary democracy inevitably legitimate the presence of protest in at least some forms. The centring of Australian political space in (this particular part of) the city of Canberra is thus a resource for regulators and protestors alike. As we have seen in recent years, the creation of any form of public dialogue about protest forms is in some ways made more difficult where the political space in question does not have such a fixed 'centre'. For example, in the face of novel protest challenges, 'global' political actors such as the World Trade Organization and the World Economic Forum have made themselves mobile, staging key meetings in a range of urban and non-urban locations which do not have any special identity as 'global' political centres – witness the World Trade Organization's move to Doha in Qatar for its 2001 meeting following the protests which successfully politicized its 1999 meeting in Seattle. As such, the legitimation of *any* form of protestatory public address directed at these institutions has been made more difficult – there are no *Guidelines* for protest at their meetings, only the promise of tear gas and truncheons.

Chapter Four

Cruising: Governing Beat Sex in Melbourne

Do you know what it is as you pass to be loved by strangers?
Do you know the talk of those turning eyeballs?
Walt Whitman, *Songs of the Open Road*, 1856

In parks and public toilets, in bars and adult cinemas, in saunas and at highway rest-stops, on beaches and online, and in countless other places, men have made contact with other men and found opportunities for sexual encounters. The practice of *cruising* for other men 'finds its proper territory in the city', according to Henning Bech (1997: 106). In comparison with our protestors from Chapter 3, perhaps cruising men seem less engaged in a form of urban public address. Sexuality is often associated with privacy rather than publicness. But in the places where men cruise for other men, the sexual intimacies which come to life have an irreducibly public dimension. Cruising, with its largely silent combinations of glances, gestures and movements, can be understood as a very particular kind of public address – the 'talk of those turning eyeballs' – which facilitates the formation of a social world shared among 'strangers'. Or, more precisely, when men cruise they are engaging in a practice which builds shared social worlds with other *men* who are both 'strangers' and 'the sort of man who takes part in the same thing as oneself' (Bech 1997: 45).[1] Through cruising, contacts are made, and 'a *being-together* is established' (Bech 1997: 113). This being-together has the characteristic circularity of a public – its form of address, cruising, simultaneously *assumes the existence of a shared world* of people who will be receptive to its 'talk' and *constructs that shared world* by enabling and sustaining the very contacts upon which it relies.

In this chapter, I explore the dynamic nature of interactions between the practices of cruising and the practices which attempt to secure the normalization of heterosexuality by analysing the regulation of 'beats' in Melbourne in the last decades of the twentieth century. 'Beat' is the name given in Australia to those non-commercial, publicly accessible places in cities when men cruise to meet other men for sex, which might take place at the beat or elsewhere. Beats are sometimes referred to as 'public sex environments' (Coxon 1996). The relationship between the counterpublic practice of cruising at beats and the normative public inscribed in the identity of 'public parks', 'public toilets' and 'public spaces' was particularly dynamic during the 1980s and 1990s in Melbourne. Over this period of time, the regulation of beats became an object of sustained and sometimes heated public debates among beat users, police, health workers, council officials, politicians and the mainstream media. The chapter asks: to what extent have these debates opened up new 'possibilities of identity, intelligibility, publics, culture and sex' (Berlant and Warner 1998: 548)? I begin by considering the particular dynamics of beat sex and the history of beats in Melbourne. While beats emerged as a response to the dominant, and often violently enforced, heterosexuality of public space in Melbourne, homosexual law reform in the 1980s changed the context in which beat sex was practised. I then consider in detail various attempts that were made to legitimate the use of public spaces as beats during the 1990s, and examine the nature of the representations of beat sex developed by advocates.

Beat-cruising as a Form of Counterpublic Address

Even the most apparently 'private' exploration of sexual desire relies on the existence of publicly accessible sexual cultures through which ideas of sexual intimacy can circulate. As Lauren Berlant and Michael Warner (1998: 547) have put it, 'there is nothing more public than privacy'. Of course, as queer theorists such as Berlant and Warner have shown, a range of public institutions, structures of understanding and practical orientations work together to normalize certain configurations of heterosexuality.[2] These configurations of heterosexuality only appear 'private' to the extent that the public support which sustains them is so taken for granted that it is almost invisible. While the privileging of heterosexuality through public institutions such as marriage and the family is relatively obvious, the normalization of heterosexuality is also 'supported and extended by acts less commonly recognized as part of sexual culture' (Berlant and Warner 1998: 555). It is:

> produced in almost every aspect of the forms and arrangements of social life: nationality, the state, and the law; commerce; medicine; and education; as well as in the conventions and affects of narrativity, romance, and other protected spaces of culture (Berlant and Warner 1998: 554–5).

The design and regulation of urban space plays its part in the normalization of heterosexuality. In modern Western cities, heterosexuality has been both presumed and reproduced 'in the designs of neighbourhoods, homes, workplaces, commercial and leisure spaces' (Knopp 1995: 154). And while the ideological privacy of sex and intimacy is mapped onto urban space, so that sexuality would appear to belong in 'private space' rather than 'public space', in fact sex is everywhere (Valentine 1993). From family photos on office desks to holding hands in a park, displays of heterosexuality across a range of apparently 'public spaces' are so ubiquitous that they are 'nearly invisible to the straight population' (Duncan 1996: 137).

Alternative sexualities, by contrast, have 'almost no institutional matrix for [their] counter-intimacies' (Berlant and Warner 1998: 562). While the circulation and reproduction of alternative sexualities relies on the existence of a publicly accessible shared world, this publicity cannot be taken for granted. Some sexualities are not inscribed in citizenship rules and welfare entitlements. The same simple acts of affection which pass unremarked when performed 'in public' by heterosexual couples (kissing, holding hands) can draw abuse and violence when performed by gays and lesbians (Mason and Tomsen 1997; Valentine 1993). The public spheres which sustain non-normative sexualities are fragile. Banished from the institutions and spaces where heterosexuality is naturalized, alternative sexualities therefore 'bear a necessary relation to a counterpublic – an indefinitely accessible world conscious of its own subordination' (Berlant and Warner 1998: 562).

While the production and regulation of urban space has certainly been complicit in the normalization of heterosexuality, the city has also been a vital resource for the construction of sexual counterpublic spheres. Increasingly in some contemporary cities, supportive spaces for non-normative sexual practices are being established through the market and local political channels, as residential and retail capital is mobilized to create urban zones which are identified with gay and lesbian 'communities' (see for example Castells 1983; Chauncey 1994; Quilley 1997; Wotherspoon 1991). In such places, non-heterosexual practices and identities may even become normalized – although the production of boundaries which can sustain these islands in the surrounding oceans of heteronormativity remains a fraught business (Moran, Skeggs et al. 2003). But in the absence of such places, and even when they are established, the creation of sexual counterpublics is 'especially dependent on ephemeral elaborations in urban space' (Berlant and Warner 1998: 562). Indeed, speaking of homosexual men, Bech (1997: 98) has gone so far as to argue that 'the city is the social world proper of the homosexual, his life space':

> Here, the homosexual can *be*; here his peculiarity can vanish in the blanket anonymity, his strangeness in the general strangeness; and here, he can make contact: either in the common urban space or in the special areas that, so to speak, concentrate the social space of the city: railway stations, urinals, parks and bath houses (Bech 1997: 98).

Of course, this is not simply a matter of the city concentrating strangers as such – the city's spaces are also *gendered* in ways that provide men with opportunities for being together with other men in the absence of women and family (Knopp 1995). Gender-segregated public toilets and sporting facilities give men legitimate reasons to be alone with other men, and provide possibilities for them to admire other men's bodies and to explore sexual attraction. Employed men also find these opportunities in gendered workplaces and leisure spaces, during work breaks and in travel to and from work. And of course, streets, parks and other 'open spaces' have historically been more hospitable for men than women – the urban '*flâneur*' who melts into the crowds in these spaces is a masculine figure. In these times and spaces, men are out of contact with family members and in an urban world shared with other men. So, while some sexualities and their associated public spheres stand in a subordinate relationship to normalized sexualities, such relationships are not static. Neither the practices of cruising nor the configurations of heterosexuality to which cruising men must relate are constant across time and space.

At beats, possibilities for homosexual attraction and encounter have been identified and exploited through the practice of cruising. Cruising at beats is a perfect example of a tactical use of urban space which sustains a form of public address (on tactics and public address, see Chapter 2). The hegemonic normalization of heterosexuality is not inverted at beats as such. Indeed, precisely because they are 'publicly accessible', the spaces used as beats are by no means set apart from webs of hetero-normative surveillance:

> The minute the homosexual gets out into town and wants to realize himself, he runs up against the police. Streets, parks, urinals, foyers, stations – all the spots where he can make contacts and, if lucky, satisfy his lusts are under surveillance, if not by the police themselves then by other guards in their place and ultimately, of course, by other onlookers (Bech 1997: 99).

Nonetheless, the normalization of heterosexuality at such spaces through surveillance is by no means totally achieved. The practice of cruising beats is calculated to exploit the opportunities some spaces provide for making contacts which do not conform to dominant heterosexual norms. Beats have typically been established in places where men could conceivably have 'legitimate' reasons for being in that place without upsetting the normative expectations sustained in these places. Some of these places are explicitly gendered and are designed to be male-only spaces (such as public toilets), while others are places where the sight of men simply walking around or hanging about is not considered 'out of place' *per se* (such as parks, road stops, beaches and the like). And the physical layouts of these places also have attributes which are put to use by cruising men – while the spaces used as beats are in some senses relatively open to strangers, they also provide cubicle doors and/or areas of dense foliage which are relatively impenetrable to surveillance. So, the practice of beat-cruising is not simply dependent on the open-accessibility of 'public space' – rather, it exploits the fact that some places are hybrids of public and private (in the realm of 'visibility' – see Chapter 1). As Smith (1993: 19) notes, 'a beat is nearly always situated in a public space which provides degrees of privacy, plus an alternative justification for being in the space other than for cruising'. The normative expectations attached to these places and their physical layouts offer cruising men a kind of camouflage. The tactics of cruising mobilize this camouflage in order to evade surveillance and punitive interventions while nonetheless making sexual contacts with other men.

Different kinds of 'gaze' are thus fundamental to beat dynamics – both the suggestive and receptive gaze which is used to make contact with other men who are interested in sex, and the surveillant gaze of urban authorities and onlookers for whom 'public' sexual contact between men is a problem (Bech 1998). As a form of counterpublic address, beat-cruising is shaped by the problem of constructing a scene of circulation which can survive in a hostile context – by making contact with those strangers who might be sympathetic (or at the very least indifferent) while avoiding contact with those who are antagonistic. Of course, negotiating the opportunities and constraints of beats for sexual contact is a very delicate procedure – and for some, the tensions involved in both achieving the kind of visibility necessary for making contact while avoiding the kind of visibility that might result in 'getting caught' considerably adds to the excitement of beat sex.[3] Here, the 'problem' of establishing a counterpublic horizon is reincorporated into the affective dimensions of the form of public address itself – a part of the experience of participating in a counterpublic which is savoured and shared.

The shared world established through cruising beats (and other places) is further reinforced through the sexual encounters which take place when men actually make contact. The being-together of men to explore sexual attraction and possibilities is 'established, consummated, and completed by means of a language, often a "silent" language, body language; moving hands, lips, and so on become signs, of liking' (Bech 1997: 114–15). This 'body language' facilitates the public circulation of knowledge about sexual possibilities that are not routinely assumed or discussed in established channels of public communication about sexuality and desire. Dowsett's (1996: 143–4) description of how men become sexually skilled through beat encounters nicely illustrates this point:

> The sexual skilling that occurs in such encounters starts with the sex acts themselves. It is about the physical possibilities of the body; what hands, mouths, penises, and anuses can achieve. A second level of skilling occurs in learning the choreography of sexual encounters. By choreography is meant the subtle and nuanced movement of bodies in sexual encounters; the stalking of partners, the shifting of attention from the general possibility of sex to the specific opportunity for sex, the inviting glance, the suggestive movements of bodies, and so on. Beyond those sensate discoveries it involves a familiarity with the context, the local sexual economy; a recognition of sites for sex as being not limited to their defined purposes and subject to certain rules of conduct.

Lest this picture of beats seems rather too static, it should be apparent that they have a geography and a history. Precisely because the practice of beat-cruising is formed in relation to the wider socio-spatial context in which beats are located, there is a dynamic interplay between beat-cruising and normative heterosexuality in which both are transformed and re-contextualized. The history of beats in Melbourne provides an illustrative example of how cruising has emerged and been constantly re-contextualized over a significant period of time.

A Brief History of Beats in Melbourne

Not long after the city of Melbourne was established, urban authorities began the construction of a network of parks and gardens. These were 'intended to improve both the moral and physical health of city dwellers', built by planners anxious to establish Melbourne as 'belonging to a network of great European cities sharing a common culture' (Whitehead 1992: 101). Of course, the public toilet facilities in these parks and elsewhere in Melbourne were gendered. The parks also included secluded spots, no doubt intended for private reflection and appreciation of nature, which provided space in which sex negotiated on the beat might take place (see Figure 4.1). These large gardens and parks have been beats for most of the twentieth century, if not longer (Carbery 1992). Fitzroy Gardens, for example, has been a popular beat for at least a century, and beat users share this public space with office workers taking their lunch break, and newly wed couples posing for photographs.

Figure 4.1   'Fern Gully' Fitzroy Gardens, Melbourne (photo: Kurt Iveson)

In Melbourne, as in other industrialized cities around the world (see for example Chauncey 1994; Delph 1978; Wotherspoon 1991), public sex between men was for most of the twentieth century both illegal and scandalous. Men who were 'caught' by police have been the subject of media attention in Melbourne since the early 1900s (Carbery 1992; Murdoch 1998). Sex between men was a criminal offence in Victoria until the 1980s, and until 1981 it was also an offence to 'loiter for homosexual purposes'. It is difficult to understand exactly how such an ill-defined concept could ever have been objectively measured by police, but as late as the 1970s this law was vigorously enforced. In the summer of 1976/77, over 100 men were charged with this offence during a police operation at one Melbourne beach alone (Carbery 1992). The repeal of this law did not bring an end to police regulation of beats. According to some activists, police continued to patrol beats as if this law was still in effect for some years after homosexual law reform in 1981 (*Melbourne Star Observer*, 7 August 1992: 4). Ten years after homosexual law reform, police in Melbourne continued to target beat users on charges of offensive behaviour in a public place, indecent exposure or indecent assault (see for example *Melbourne Star Observer*, 7 February 1992: 6; *Brother Sister*, 1 May 1992: 3). In response to complaints about men loitering around parks or public toilets – and according to activists, sometimes for their own amusement – the Victoria Police have routinely conducted 'entrapment' operations at beats. In these operations, sexy young police officers in plain clothes encouraged other men to think that they were interested in a sexual encounter, and then charged these men with either offensive behaviour or indecent exposure if they exposed themselves, or with the more serious offence of indecent assault if they touched the police officer. These charges were (and are) often not contested by beat users – to do so would involve a court appearance, and risk of exposure in the press or directly to family or employers (Police Lesbian and Gay Liaison Committee (Vic) 1996: 23). Police operations at beats thus result in:

> easy arrests and convictions, because most men charged with beat offences plead guilty, whether gay or not, whether they actually did anything or not, just to 'keep their names out of the papers', as the Police menacingly put it (*Melbourne Star Observer*, 7 August 1992: 4).

Police are not the only ones who have attempted to curtail beat sex. Fear of exposure has been manipulated by men posing as police and demanding money from beat users, in a scam which caused as much concern to police as to beat users (*Brother Sister*, 4 June 1993: 5). The pleasures of beat sex are also decidedly unpopular with those who provide the publicly accessible facilities that are used. Along with intravenous drug use, 'loitering by men seeking homosexual contact' was identified by Melbourne City Council in a 1997 report as one of the main security concerns to be addressed in the design and location of public toilet facilities (Melbourne City Council 1997). Beat users also risk violent bashings and even murder from gangs of men stalking beats (Swivel 1991; *Brother Sister*, 27 May 1999: 3). Indeed, as Smith (1993: 19) has noted:

> Traditionally, all the interested parties of beats, apart from beat users themselves, have served as repressive agents devoted to eliminating or at least regulating the existence of beats.

Reshaping the Public World of Beats in Melbourne: Making Space for 'Safe' and 'Legal' Beat Sex

The 1980s and 1990s witnessed two major sets of interventions in beats which reshaped the possibilities for beat-cruising in Melbourne. These interventions are significant because they contested traditional regulatory attempts to eliminate beat sex, seeking instead to engage the counterpublic world of cruising men in order to promote 'safe' and 'legal' beat sex. As we shall see, while the institutional actors promoting 'safe' and 'legal' beat sex were more sympathetic to the practice of beat-cruising, they also had negotiate a series of tensions that arose when they sought to publicly represent and debate beat practices which had until then survived only by *avoiding* such a widening of public knowledge of their particular dynamics and styles of 'talk'.

Safe beat sex: the Beats Outreach Project

From the 1980s onwards, the governance of beat-cruising was reshaped through the development of policy responses to the HIV/AIDS pandemic. Since this time, gay and lesbian community organizations have been funded by the Australian Commonwealth Government to provide information about safe sex to gays and lesbians as part of the *National HIV/AIDS Strategy*. This arrangement had been established on the grounds that information provided by these organizations would be more relevant and accessible to their target audience. By the late 1980s, however, it had become apparent that while such strategies were having some success, not all men who engaged in homosexual sex would relate to this information, particularly if they did not identify with the 'gay community':

> In community based education efforts, especially those devised by community based organizations for use in the gay community, the idea of community norms and a safe sex culture has become the dominant perspective on achieving and sustaining behaviour change . . . [T]he community norms and community attachment model which has been applied with gay men becomes problematic in relation to men who place themselves outside the notion of a gay community (Bartos, McLeod et al. 1993: 13, 16).

Men who did not attach themselves to the gay community but who engaged in homosexual sex came to be referred to as 'men who have sex with men' or 'MSMs'. This categorization, as noted in a report to the Department of Health about the meanings of sex between men, was controversial. Some feared that it was an attempt to displace the role of gay community organizations in public health provision to homosexually active men (Bartos, McLeod et al. 1993: 2). But targeting MSMs presented a more fundamental challenge to public health campaigns, because as a 'group' MSMs had no shared culture or identity:

> To talk of MSM as a definable 'group' is really a misnomer. As a conceptual category for purposes of research, planning and health promotion it has some use. However, it is not useful to think about these men as a group if we are looking at their sexual experiences and how they fit them in to the pattern of their lives (Bartos, McLeod et al. 1993: 69).

In other words, the problem with existing health promotion strategies was not simply that they targeted one group or identity ('gay') instead of another ('MSM'), but rather that the very notion of targeting groups could only ever work for people who identified with a group. Targeting, as it existed, was 'predicated on the prior existence of defined groups to which people see themselves belonging' (Bartos, McLeod et al. 1993: 72).

To deal with the limitations of existing community-based models of health promotion, new strategies were designed with the intention of 'engaging' men in relation to different *behaviours* or *situations*, rather than community identifications. As advocates of engagement put it:

> rather than seeking closer targeting on populations which are specified in greater and greater detail, the task of health promotion can be seen as seeking an active and inclusive engagement of its audience. The notion of engagement still allows for particular health promotion messages to be devised in relation to different behaviours or circumstances, but there is less attention to the characteristics of the relevant population (Bartos, McLeod et al. 1993: 74; see also Dowsett 1996: 72–3).

Engagement was not a universalist means of public health provision – it still paid attention to specific characteristics of target groups, but these target groups were to be defined in a situation-specific way rather than on the basis of a shared identity or culture. In the United States, Cindy Patton (1991: 152) has considered such strategies in terms of their ability to understand and work with established 'sexual vernaculars':

> For people who are not part of a self-identified sexual community, you have to find out how they negotiate sex, how, for example, men who have sex at a truck stop find out that the truck stop exists as a place for sex . . . Bodies somehow manage to connect, people manage to have sex, and I think that there are regular or, perhaps, irregular rules for finding people who will engage in those activities with you. And this is what I mean by sexual vernacular.

Patton's notion of a 'sexual vernacular' here is very similar to what I have referred to as a style or form of public address associated with a particular (counter)public sphere. As a model of health promotion, then, engagement relied upon the ability of health promotion agencies to tap into the counterpublic spheres through which these alternative sexual vernaculars were developed and sustained. As such, the sexual public which took shape through beat encounters was now to be drawn upon as a *resource* for governmental interventions in sexual practice seeking to 'find their audience', re-imagined here as MSMs.

In Australia, the notion of engagement drew inspiration from a number of health promotion projects established by State-based AIDS Councils that targeted beats. These projects provided safe sex material and information at beats, and were supported by Health Departments as a way of accessing MSMs as well as gay men. The threat of some men picking up HIV at a beat and 'taking it home' to their heterosexual partners was of particular concern to the government agencies which funded these projects (Smith 1993). Early attempts to establish beat outreach projects in the late 1980s identified the importance of 'systematic observation and documentation of beat use and beat user practices – beat mapping' (Dowsett and Davis 1992: 25). Beats outreach workers, in other words, had to develop a familiarity with the sexual vernacular of beats in their area of operation, as well as keeping in touch with the shifting geography of beats. To assist in this process of mapping, a considerable amount of research was conducted on the nature of beats and beat sex, addressing questions such as: who used beats? how do they vary geographically? what kind of sex do men engage in at beats? (see for example Dowsett and Davis 1992; Bartos, McLeod et al. 1993; Smith 1993; Fraser 1995). The results of this research were presented in health journals, academic monographs, conferences, and in reports funded through the *National HIV/AIDS Strategy*, to assist in the ongoing development of public health engagement strategies. Of course, cruising and public sex were not new objects of the social-scientific gaze, although earlier research efforts had been framed rather differently as attempts to understand the sociology and psychology of sexual 'deviance' (Delph 1978). While contemporary 'beat mapping' efforts might have differed in significant respects from this earlier research, they had in common the effect of circulating new knowledge about beats, in the process helping to expose the beat vernacular to a wider audience.

Drawing on this research and on the experience of early outreach projects, a growing number of health authorities and community-based agencies in Australia (Dowsett and Davis 1992; Baird 1997) and internationally (Woodhead 1995; Brown 1999) began to engage with men using beats in the provision of safe sex education. In Melbourne, the Victorian AIDS Council/Gay Men's Health Centre (VAC/GMHC) established a Beats Outreach Project (hereafter 'Beats Project') in 1990. The Melbourne Beats Project consisted of one paid worker, and a number of trained volunteers drawn from the population of beat users who were familiar with beat culture. As one Beats Project worker told a gay and lesbian weekly newspaper in 1994:

> These guys are aware of the etiquette of beat cruising. They are sensitive to the gestures, the eye contact, the pace and the attitude many men present at the beat (*Melbourne Star Observer*, 15 April 1994: 4).

Like most other outreach projects, the Beats Project in Melbourne was informed by the concept of harm minimization – the pleasures of beat sex

Figure 4.2   Pocket-sized safe sex booklet distributed by the Beats Project

were recognized, and the aim of the project was to encourage safe sex among beat users to prevent the spread of HIV, rather than to curtail beat sex.[4] In Melbourne, through the use of volunteers, public health information was provided at the beat itself in ways it was hoped would not alienate beat users. Pocket-sized information booklets were produced in recognition of the need for disposable and easily portable material (see Figure 4.2). Condoms and lube were left in lit buckets at beats for beat users to take discreetly, in recognition of the non-verbal cues which for the most part constitute beat culture. The Beats Project produced a small bulletin, the *Ugly Punter*, which offered information about violence at beats, descriptions of offenders, advice to those threatened at beats, and points of contact and support for those who were assaulted at beats (*Brother Sister*, 9 January 1997: 9). The VAC/GMHC also produced stickers about the Beats Outreach Project and safe sex, which were stuck on toilet doors and walls in known beats.

Although it was recognized that not all beat users in Melbourne identified as 'gay', the VAC/GMHC nonetheless found that many accessed the gay press which was used to both promote the Beats Project and provide information about safe beat use. Over the 1990s, a number of feature articles written by project workers appeared in both *Brother Sister* (for example

31 December 1993: 3; 22 April 1994: 10; 2 November 1995: 13; 11 July 1996: 11) and the *Melbourne Star Observer* (for example 7 February 1992: 6; 3 April 1992: 6; 15 April 1994: 4). Advice about safe beat use was not restricted to the transmission of HIV and other sexually transmitted diseases, but also included advice about how to avoid detection, and thus potential violence and prosecution. For example, the article 'Beats – how to do them safely', written by VAC/GMHC Outreach Officer in the *Melbourne Star Observer* (3 April 1992: 6), had a range of tips for potential beat users such as:

- become familiar with the spaces of the beat before trying it out;
- 'to give potential perpetrators of violence or harassment the wrong impression, camouflage your real intent for being there: ride a bike, take "Blanche" the dog for a walk, dress as though you were jogging. There are lots of reasons for being around the area other than for sex';
- carry condoms and lube;
- 'if you are approached by the police don't panic, you have rights . . . If you have camouflaged your intent for being at the beat, say you're there for a jog or to walk the dog or whatever; use your imagination. For the police to charge you, you must be caught in the act of committing offensive behaviour in a public place'; and
- report acts of violence to 'gay sympathetic services' such as the Beats Project or the Police Lesbian and Gay Liaison Committee, if not the police themselves.

Project workers faced at least four difficult issues in their efforts to promote safe beat sex by engaging with the beat-cruising vernacular. First, the project had to secure some base of funding. In fighting for a greater awareness of safe beat use, the VAC/GMHC had to overcome the fact that men meeting other men for sex in public places remained 'the stuff of the moralist's worse nightmare' (Dowsett and Davis 1992: 5). While beats outreach projects did have varying levels of government support in different jurisdictions, such support was low profile:

> none would defend the projects too vigorously, for fear of criticism related more to moral concerns about such transgressive sexual activity than to the value of and necessity for innovative outreach interventions such as these projects (Dowsett and Davis 1992: 18).

Second, according to some project workers, beats were disparaged by some within the established gay community as well as by more mainstream 'moralists'. According to one worker on the Beats Project, there existed a:

gay hierarchy of glamour and acceptability which places dinner, dance and tupperware parties at the top; clubs and bars in the middle; saunas and other sex-on-premises venues down the lower end, and beats right at the bottom. The image of beats as the domain of married, olds, desperates and poor dancers is often, unfortunately, held by beats users themselves (*Brother Sister*, 2 November 1995: 13).

Nonetheless, he argued that because of the very existence of such a hierarchy and the commercialization of gay culture, beats would not go away. Gary Dowsett's (1996) study of homosexually active men in Sydney also found that beats act as an alternative to the established commercial gay 'scene' for men who find it either inaccessible or alienating. Safe beat sex, then, posed a challenge to the established gay community as well as to the normative 'public' inscribed in the 'public spaces' of Melbourne.

The third issue raised by the engagement model of health promotion, and a particularly interesting issue for my purposes here, concerned the impact of the Beats Project on beat dynamics. By heightening the visibility of beat-cruising practices, the Beats Project potentially endangered the conditions that sustained the very sexual vernacular with which they sought to engage. Reporting of research in policy, academic and community media outlets circulated knowledge about beat dynamics to a wider audience. Perhaps more significantly, safe sex campaigns made beat-cruising more visible 'on site'. One of the unintended consequences of safe sex campaigns at beats was an increase in litter and promotional material, which made the occurrence of sex in publicly accessible places much more visible to other users of those sites. The horizon of the sexual counterpublic which took shape through beats became more difficult to manage in this context. Condoms and stickers drew attention to the use of parks and public toilets as beats and threatened the discretion that is fundamental to beat-cruising. VAC/GMHC spokespeople asked beat users to 'use it and bin it', citing instances of:

> council workers, cleaners and gardeners actively frustrating beat users and other gay men by alerting employers or the police to used condoms lying around. In some cases beats have been locked up after council workers complained at having to remove used condoms every morning (*Brother Sister*, 31 December 1993: 3)

A letter to *Brother Sister* (14 January 1994: 8) from one beat user noted that 'A clean beat is a still open beat':

> I can't believe these gay men. They get upset when council starts locking toilets of a night time, and cutting back trees at beats. If the men cleaned up

after themselves, toilets might still be open, and trees and bushes might still be growing.

Finally, in attempting to establish an institutional involvement in beats that did not actively discourage beat sex, the Beats Project was drawn into relationships with institutions like the police and local government whose practices already helped to shape the wider institutional matrix in which beat sex took place. Project workers spent a good deal of time negotiating with police about the regulation of beats in order to prevent their own arrest or harassment, and to ensure that police operations did not disrupt outreach activities. They feared that they would be seen as *de facto* police if beats outreach coincided with police operations – a situation which would quickly see beat users lose trust with them. In order to address this issue, project workers attempted to build relationships with police on a local, personal scale. At the same time, however, the operational strategies used to police beats became caught up in wider public debates about the policies of the Victoria Police towards gays and lesbians. While health promotion workers sought to intervene in beat-cruising practices in order to promote *safe* beat sex, other activists attempted to establish policing methods which recognized the *legality* of beat sex. I now want to turn to these efforts, to further explore the impact of institutional interventions on beat-cruising in Melbourne.

Legal beat sex: the Police Lesbian and Gay Liaison Committee

Police treatment of gays and lesbians had been the subject of sustained activism by gay and lesbian advocates in Victoria since the 1970s. This activism focused for the most part on three related issues: the number of gays and lesbians in the Victoria Police; the harassment of gays and lesbians by police; and the accessibility of the Victoria Police to gays and lesbians who are victims of crime. One of the main avenues for this activism was the Police Lesbian and Gay Liaison Committee (PLGLC). This committee was established after homosexual law reform in Victoria in 1981, and was reactivated in the late 1980s following a period of inactivity. Members of the PLGLC were elected at public meetings advertised in the lesbian and gay press and community networks. The purpose of the PLGLC was to advise the Victoria Police on matters pertaining to the policing of lesbians and gays. To this end, it established a relationship with the Community Liaison branch of the Victoria Police, with whom informal and formal meetings were held regularly. The PLGLC also raised its profile in Melbourne's gay and lesbian community. Gays and lesbians

Figure 4.3   Police surveillance notice, mens' toilets, Fitzroy Gardens (photo: Kurt Iveson)

were encouraged to treat the PLGLC as an avenue through which they could make complaints about police conduct, if they did not feel they could raise these matters with police directly for fear of harassment or intimidation.

Initially, the PLGLC focused much of its energy on changing negative perceptions of gays and lesbians, which it felt were entrenched within police culture. Members of the PLGLC presented information on gay and lesbian sexuality to police recruits and junior officers at the police training academy. Beats were one of the most confronting issues for police participants in these workshops. Peter Horsley, the PLGLC member who spoke at the initial police seminars, noted that 'most police interest centred on questions about men who had sex at beats or toilet blocks, and fears over dealing with HIV positive people' (*Brother Sister*, 15 May 1992: 5). This police concern about beats was not surprising, and was reflected in the frequent complaints to the PLGLC about police conduct of entrapment and decoy operations (see Figure 4.3).

The policing of beats became a key issue for the PLGLC as the 1990s proceeded.[5] Members of the PLGLC made referrals to legal and support services for men charged with beat-related offences, and provided court support. The PLGLC also seized on the decision of a Melbourne magistrate in a 1992 case involving a man who was charged with offensive behaviour in a public place, resisting police and escaping lawful custody after an encounter with police while cruising.[6] The defendant in the case admitted that he had met another man at a beat in a park in inner-city Melbourne at around 12.30 a.m., and that they enjoyed some mutual masturbation and oral sex under the cover of a dense tract of bushes. He alleged that on leaving the bushes, he was approached by two other men, who later identified themselves as plain-clothed police officers, and wrestled to the ground. He also alleged that when he complained of his treatment, he was called a 'smart cunt' and received a few hard kicks to the groin and ribs. The man contested the charges based on a legal precedent set in a 1961 Supreme Court case (*Inglis v Fish*) which found that behaviour could not be considered offensive if witnessing it required abnormal or unusual actions. In this case, the magistrate found that the fact that the sex occurred in a secluded place in the middle of the night required unusual actions to witness it. The charges were dismissed, and the man was awarded legal costs.

In beating the charges, the defendant had manipulated the complex positioning of beat sex in relation to conventional topographical mappings of the public/private distinction. As Law et al. (1999: 53) have noted in another context:

Because using public toilets is simultaneously defined as both intensely private and of public concern, the history of claims for access to public toilets reveals the complex intertwining of public and private over time.

While a headline following the case in the *Melbourne Star Observer* declared 'Court backs public sex' (*Brother Sister*, 24 July 1992: 3), in fact the *privacy* of beat sex was used to protect, even legitimate, the ongoing practice of an alternative *public* sex culture – for the purposes of legislation pertaining to offensive behaviour, the time–space of actions which took place in the middle of the night in a heavily secluded spot were deemed not to qualify as 'public'. The case was widely reported in the gay press, along with guides about the legal rights of beat users. Other men were urged to challenge police charges (*Melbourne Star Observer*, 3 April 1992: 6; 7 August 1992: 4; 17 September 1993: 4; *Brother Sister*, 24 July 1992: 3). Coverage of the case ensured that a locked cubicle door came to be widely accepted by activists and police as a barrier which protects men from the charge of 'offensive behaviour in a public place'. According to Jamie Gardiner, a long-time gay activist:

This victory does not mean that gay men can go around having sex in public wherever we feel like it. But it does mean that discriminatory policing and prosecuting need not be tolerated and will not win in the courts (*Melbourne Star Observer*, 7 August 1992: 4).

In the years immediately following this case, some progress was made in shifting policing operations at beats in Melbourne. In 1993, PLGLC volunteers were invited to participate in a combined operation with police in South Melbourne, 'Operation BusyBeat'. This involved members of the PLGLC joining two uniformed police officers in patrolling a beat that had been the subject of complaints from nearby residents. No one was to be charged with beat offences during the week-long operation. Rather, its purpose was to inform beats users of the law regarding beat sex through a highly visible operation, as an alternative to entrapment. Members of the PLGLC handed out leaflets informing users of what was legal and what was not. The leaflet stated that:

It is not illegal to come to this place or meet people here. However, Police can take action where the following offences occur: offensive behaviour; indecent assault; indecent exposure; loitering for prostitution (Police Lesbian and Gay Liaison Committee 1994: 8).

The PLGLC thought the operation a great success, and believed it provided a useful model on which new police policies and procedures could be based:

Plain-clothes operations, which often create crimes by individuals (decoy entrapment) should be a thing of the past. Modern methods of crime prevention such as [Operation BusyBeat] are a far more enlightened way of preventing crime and responding to public concerns, rather than creating crime and playing into the hands of homophobic police who delight in ruining the lives of gay, bisexual and married men who have sex with men at beats (Police Lesbian and Gay Liaison Committee 1994: 5).

Over the next two years, the PLGLC focused much of its attention on developing a formal proposal for a beat-policing policy for adoption by the Victoria Police which would recognize the legality of men meeting other men at beats. Early in 1994, discussions between the PLGLC and a newly appointed Gay and Lesbian Liaison Officer at police headquarters, Chief Inspector John Winther, resulted in the development of an informal protocol for the policing of beats. Winther wrote to local Patrol Commanders asking them to give him notice of any police operations at beats, so that he could pass on information to the PLGLC:

In doing so the Gay Community can be advised that police patrols of certain areas will occur and action will be taken against persons found offending. I believe that this will have a two fold effect, firstly the Gay Community will be proactively warned and secondly, the possible reduction in the duration of the operation and personnel deployed (Winther 1994).

This practice, although not a formal requirement for Patrol Commanders, resulted in the gay press publishing notice of police operations at specific beats (see for example *Melbourne Star Observer*, 2 September 1994: 3; 24 January 1997: 3; 19 September 1997: 3). But as one Beats Project worker noted, 'the main concern about this process is that this information does not reach those men who do not identify as gay and do not read gay newspapers' (*Brother Sister*, 2 November 1992: 13).

Efforts to reform beat policing in Victoria were given further impetus in 1995, after a formal protocol for the policing of beats was adopted by police in the neighbouring state of New South Wales. The NSW Police Commissioner's Circular concerning beats was the culmination of four years of work by a Taskforce involving both police and representatives from the AIDS Council of NSW (ACON), and the instructions in the circular were updated and included in the 1999 *Police Service Handbook*. When targeting offensive behaviour at beats, police were instructed to prioritize the use of marked police vehicles and uniformed police. Covert operations were only to be conducted with the written approval of patrol commanders, who in turn were instructed only to approve such operations in consultation with Gay/Lesbian Liaison Officers, and other interested parties such as ACON beats workers. Such covert operations were thought appropriate only in investigations of complaints of assaults at beats. The Commissioner expressed the hope that the new protocol would reduce accusations of entrapment and police misconduct at beats, thus having even wider and more positive ramifications:

An important consequence of professional police conduct when responding to complaints regarding offensive behaviour at 'beats' will be an increase in the level of confidence and trust in reporting crimes of violence, and the subsequent apprehension of assailants who might otherwise go undetected (New South Wales Police Service 1995).

Working towards a similar policy, in December 1996 the PLGLC presented a report to the Victoria Police about beat policing – *Policing public place 'beat' meeting behaviour – recommendations for the Victoria Police*. This report presented information about beat behaviour, and recommended the implementation of a formalized police policy on beats. According to the PLGLC, there were five principle reasons for such a policy:

1. to reduce the instance of criminal offences associated with beats, especially violence;
2. to address the consequences of potential disclosure/outing with the required sensitivity;
3. to enhance police integrity and reduce allegations of police misconduct;
4. to increase cooperation with other harm reduction initiatives and education;
5. to uphold individual rights of assembly and association (Police Lesbian and Gay Liaison Committee (Vic) 1996: 11).

The report reinforced the notion that beat sex taking place out of the public view was not criminal, and urged police to acknowledge that 'men congregating in public places for homosexual purposes were engaging in lawful activity' (1996: 7). While the report acknowledged that sex taking place in full public view was illegal, it emphasized that sexual acts at beats 'by and large go unnoticed by the community until the entry of the police as a third party' (1996: 2) – in other words, offensive behaviour or indecent exposure only occurs when encouraged by undercover police officers for the purpose of making arrests. The report documented a police operation in Carlton Gardens which had resulted in 16 prosecutions, not one of which had involved another party other than a police officer in plain clothes (1996: 46). Based on an explanation of beat dynamics and history, it was argued that the existing culture of beat use was discreet, and 'designed not to offend the public but solicit interest only from those familiar with its use' (1996: 15). While beats were 'publicly accessible' places, the report distinguished between the 'public' activity of soliciting sex, and the 'private' activity of engaging in (various kinds of) sex (1996: 16).

The report did not request the complete withdrawal of police from beats, but rather attempted to recast the terms on which such police involvement could be justified. The PLGLC's priority was to stop police entrapment operations, as well as other police harassment of beat users such as the issuing of parking infringements at deserted beats after midnight (see for example *Brother Sister*, 13 August 1993: 8), and even the use of police helicopters to scatter men in parks and beach scrub at night (Police Lesbian and Gay Liaison Committee (Vic) 1996: 37–8). The PLGLC report encouraged police to be responsive to complaints about violence on beats, and to engage in uniformed operations only when requests for surveillance were made by members of the public. Such complaints were to be put in their proper legal perspective:

Homosexual law reform did not simply replace the law of 'loitering in public places for homosexual purposes' with 'offensive behaviour' or 'soliciting for prostitution'. It accepted homosexuality as being lawful. It accepted that men congregating in public places for homosexual purposes were engaging in lawful activity (1996: 7).

The ongoing education of beat users in safe sex by the Beats Project was used to reinforce the harm minimization message of the 1996 PLGLC report. Insensitive police operations, it was claimed, prevented the Project from functioning effectively. Beats Project workers consulted by the PLGLC identified three problems with the existing plain-clothes policing of beats:

- Beat users leave condoms and lube at home, or do not carry safe sex materials with them, for fear of identification by police at beats;
- Beat users mistake outreach workers for undercover police operatives and the opportunity to educate these individuals may be lost, and further educational interventions with these people may be compromised. This problem is even greater when hard to reach clients such as former prisoners, youth and the intellectually disadvantaged, are involved;
- Clients will change their meeting places to places unknown after police activity focussing on particular beats. The educational opportunity is often lost until the alternate meeting place is identified (1996: 48–50).

Further, outreach workers reported difficulties with specific patrol areas (such as Geelong), where police had failed to respond to requests of assistance by workers fearing assault either on themselves or other beat users, thus 'resulting in loss of client-worker confidence, loss of confidence in police back-up for health educators working in the field, and cessation of outreach activities' (1996: 50).

The report revealed some ongoing matters of dispute between police and the civilian members of the PLGLC about the best way forward. One proposal for a new protocol developed by Chief Inspector Winther was reprinted in the report, with a note about its rejection by the PLGLC (1996: 51–2). It was followed directly by the model adopted in NSW (1996: 53–5), in a comparison designed to draw attention to the shortcomings of Winther's proposal, which was in the form of suggested guidelines for policing beats rather than operational requirements. Nonetheless, an emerging mood of goodwill seemed to bode well for a favourable response to the report's recommendations. The police were now publicly campaigning to increase the reporting of homophobic violence. Dennis Cairns, Superintendent in charge of Community Liaison, told The Age (1 January 1997: 3) that lesbians and gays should 'be reassured that if they reported assaults to police they would receive understanding from officers, and complaints would be pursued'. The coordinator of the Beats Project approved: 'The police need to be applauded for coming out like this, and taking a proactive role to stop (attacks). It's just fantastic'. The Secretary of the PLGLC, Dolf Boek, also applauded the police's action in calling for gays and lesbians to report violence. Not everyone was convinced of the police's good intentions, however. *Brother Sister* supported the apparent shift in attitudes towards gays and lesbians, but the editor asked:

> why does Victoria still have only one Gay and Lesbian Liaison Officer, who also liaises with two other minority groups, while NSW has over one hundred? And, why has most of his work with our community been only to warn us of which beats are being policed in an undercover operation? Is this their idea of proactive (*Brother Sister*, 9 January 1997: 12)?

These lingering doubts about the Victoria Police were soon confirmed. One month later, the Minister for Police Bill McGrath told the ABC television's *7.30 Report*[7] that he had 'great reservations about gays and lesbians in our police force' (*The Age*, 20 February 1997: 1). This provoked an immediate and angry response from gay and lesbian organizations, and members of the Gay and Lesbian Police Employees Network (GALPEN) expressed disappointment in the Minister's comments (*The Age*, 20 February 1997: 1; 21 February 1997: 2). The existence of this network of gays and lesbians within the police force, according to Joseph O'Reilly from the Victorian Council for Civil Liberties, had:

> enhanced the faith of gay and lesbian citizens in the capacity of the Victoria Police to respond to them with a modicum of sensitivity. In the gay and lesbian community, the network was seen as an antidote to the force's inherent homophobia. It provided one way of redressing police harassment and discrimination – evidenced in the Tasty nightclub raid[8] – with which most gay men and lesbians associate the police (*The Age*, 21 February 1997: 17).

At first it was hoped that McGrath's comments as Minister would have little effect on the relationship between operational police and the PLGLC – Boek told the *Melbourne Star Observer* (21 March 1997) on behalf of the PLGLC that 'We don't think his ignorance will make much difference'. However, in September 1997, Victoria Police took disciplinary action against GALPEN spokespeople who had spoken out against McGrath's comments. Boek publicly condemned this action, and the PLGLC voted to suspend Chief Inspector Winther, accusing him of obstructing their attempts to examine the disciplinary action taken against GALPEN members. The police responded by withdrawing their support for the PLGLC altogether. Superintendent Cairns justified this action by arguing that the PLGLC had pursued 'non-relevant' matters, and said 'the Force will concentrate its efforts on liaising directly with members of the gay community who remain interested in genuine cooperation and constructive dialogue' (*Brother Sister*, 2 October 1997: 3; see also *The Age*, 24 September 1997: 9). Both the secretary of the PLGLC and Joseph O'Reilly described relations between the gay and lesbian community and the police as being at a ten-year low. Hot on the heels of this dispute came the publication of the PLGLC's survey of 325 lesbians and gays (conducted at the annual Midsumma Festival), which found that of those who had experienced personal/property crime or victimization/harassment during the previous two years, 77 per cent chose not to report the incident to police. Many of these respondents did not report incidents to police on the grounds that they felt their report would not be addressed seriously and empathically by police (*Brother Sister*, 30 October 1997: 5).

The collapse of relations between the PLGLC and Victoria Police put an end to negotiations on new beat policing policies, and the 1996 report's recommendations were never adopted. In the absence of a State-wide policy on beat policing, arrangements for the regulation of spaces used as beats remained locally specific, fluid, and often susceptible to the personality and attitudes of local police officers. Some of these attitudes stubbornly refused to change. Even after his involvement in lengthy discussions on the subject in the early 1990s, Police Superintendent Cairns said that there was only so much that police could (or would?) do about the safety of beats:

> The answer lies to a large degree with the gay community itself. Toilets are designed to be toilets and not for men to have sex in (*The Age*, 29 November 1997: Good Weekend, 18).

For Cairns, *no* beat sex was preferable to safe beat sex. Some beat users agreed with Superintendent Cairns that the answer to their problems lay with them, although their answer was somewhat different to the one he had in mind. In the face of a sustained campaign of homophobic intimidation and attacks at beats in Geelong, beat users formed a group of vigilantes. They penned threatening notes to suspected bashers, and slashed their car tyres. Their action eventually attracted the attention of police to ongoing problems in the area, where earlier complaints of violence had failed to draw a response (*Brother Sister*, 6 March 1997: 5; 20 March 1997: 3). Not long afterwards, a Geelong police officer, possibly disgruntled by the actions of the vigilantes, provided the *Geelong Independent* with a list of beats in the area, which they promptly displayed prominently on their front page (21 February 1997). He reassured the paper that 'we are certainly aware that there are certain areas that homosexuals frequent, and we keep files on paedophiles and unusual sexual behaviour'.

Conclusion: The Risks of Cruising and Claim-Making

The history of interventions in beats in Melbourne that I have discussed in this chapter is a story of various efforts to reshape the opportunities afforded by various places in the city for sexual contact between men. As such, it constitutes an excellent example of a struggle over the urban dimensions of counterpublic-making. As a form of public address, cruising at beats is both the prerequisite and the product of a sexual counterpublic sphere which is fragile and unstable. This fragility and instability is the result of the wider conditions in which beat-cruising takes place. The ongoing existence of a sexual counterpublic of beat users relies on the capacity of men to make sexual contact with other men – but the form that such contacts take is conditioned by an awareness that not all men are the kind of men who would welcome sexual advances from other men, and that being identified as a man engaged in cruising could have serious social, legal and physical consequences. In Melbourne, as in other cities, a form of public address which meets these requirements has taken shape as some men have tactically mobilized the opportunities afforded by particular urban spaces for homosexual contact. This is not simply to say that cities provide opportunities for cruising – it is also to say that cruising is itself a form of public address which is fundamentally urban. 'The city', as Bech (1997: 118) has argued, 'is not merely a stage on which a pre-existing, pre-constructed sexuality is displayed and acted out; it is also a space where sexuality is generated.'

As things currently stand, after two decades of concerted struggle, beat sex in Melbourne takes place in a new context. The engagement efforts of safe sex workers, whose interventions in beats are informed by the notion of harm minimization, are well established. The Beats Project continues to promote the notion of safe beat sex, and there has also been some clarification of what constitutes legal beat sex. However, the interventions of project workers and activists exist alongside continued repressive involvement from other institutions. Where possible, parks are increasingly locked at night. Some public toilet facilities have been redesigned to single, unisex facilities with time-releases on the doors and no vestibule areas (which apparently 'tend to be used to shield people engaging in criminal or anti-social behaviour'), and new facilities are placed in well-lit areas on street corners, the edge of parks, and even median strips on roads (Melbourne City Council 1997: 3). And while some police are supportive of the notions of safe and legal beat sex, others clearly continue to be mystified by some men's taste for cruising and actively seek to stamp it out.

Despite the significant differences between sympathetic and hostile interventions, there are also some important similarities in the nature of these interventions. In their own ways, health workers, gay and lesbian activists, police and council workers each sought to generate knowledge about the sexual vernacular of beat-cruising which could be put to work in support of their agendas. An understanding of this sexual vernacular was fundamental both to police engaged in entrapment operations and health promotion workers trying to encourage safe sex at beats. This knowledge was generated and circulated through a range of fora and media in which beats came to be an *object of public address and debate* – from observations at beats themselves to health policy conferences and medical journals, court rooms, the gay and lesbian press, police consultative committees and local government technical committees, and occasionally mainstream local and mass media. The debates in these fora and media concerned both the identities of the spaces used as beats and the identities of men who cruise beats – so, for example, the interventions of health workers at beats were calculated to engage with 'men who have sex with men' or 'MSMs' in 'public sex environments', the interventions of activists were calculated to clarify and secure the rights of 'homosexual men' and 'gays' by clarifying when and where they were 'in public' or 'in private', and the interventions of police and council workers were calculated to shut down beats in order to secure the spaces used by 'the community' or 'the public' against 'deviants' and those engaged in 'anti-social behaviour'.

The generation and circulation of knowledge about beat-cruising, then, is mobilized in ways that have potentially significant impacts on the context in which it is practised. Even the more sympathetic institutional interventions can heighten the visibility of beat sex in ways that compromise the delicate balance of visibility and invisibility which defines the sexual vernacular of cruising. As a consequence, there are risks involved in the wider circulation of the 'sexual vernacular' of the cruising counterpublic:

> people from certain subgroups become afraid to speak their native tongue when their 'texts' – a red hanky, a term of phrase or a cut of suit, a pamphlet, a book – thought private, suddenly come under scrutiny and become public, rendering the private language and symbols of the subculture vulnerable to unanticipated readings by someone with greater social power. And, on the other side, members of dominant language communities feel their territory has been invaded with languages they do not wish to acquire (perhaps because those languages highlight, perhaps for the first time, the experience of the irregularity of the borders and their arbitrary and unjustifiable concentration of power) (Patton 1991: 47).[9]

Given these risks, such interventions have had their critics. David Woodhead (1995: 239) has argued that the apparent 'privacy' promised by the cubicle door has never been stable, and worried that the tactics of invisibility used by cruising men were fragile enough without the involvement of beats workers:

> Despite the exclusionary promise of the cubicle door, the world is never wholly shut away. The open top allows the public and reminders of the public world (vehicle noise, people talking) into the cubicle. Most importantly, it allows easy access for those whose interest in what is happening therein is less concerned with personal sexual activity and more with disrupting those alleged practices.

The threat to cruising conditions constituted by safe sex projects has occasionally even drawn a physical response, with outreach workers being chased away from beats by angry men who do not want to be 'engaged' (Joseph 1997: 85). In a twist on Foucault's notion of 'the gaze', Woodhead refers to safe sex educators on beats as 'surveillant gays'. He draws the conclusion that public health projects, by exposing beat users to increased institutional involvement from outreach workers and others, threaten the very nature of beat sex as a practice resistant to heteronormativity:

> To recognize the value of spatiality, to formally assert its importance in the constitution and expression of identities, and then to expediently appropriate those spaces as sites to dispense disciplining knowledges, is to initiate and enact a series of practices that can result in disciplining the actions of those individuals whose resistances are often beyond the reach of the state (Woodhead 1995: 243).

To a certain degree, the health promotion workers attempting to learn the sexual vernacular of beat sex in order to make it 'safer' take their place in a longer history of medicalized interventions in the lives of men who have sex with other men which are purportedly of a different order to police interventions, but which nonetheless are concerned with discipline and governance. For the homosexual man:

> Whereas the police are a danger and a limit he attempts to avoid (or perhaps nuzzles his body surface against), the physician is allegedly a friend, a confidant. He (sic) can safely be told everything, for his only wish is to help; and he needs to know everything, or he cannot help (Bech 1997: 101).

This critique does point towards some of the serious implications of engagement. But the notion that sex at beats could be 'beyond the reach of the state' misrecognizes the position of the beat-cruising sexual counterpublic with respect to the normative public inscribed in the spaces through which it takes shape. While the form which cruising takes is calculated to minimize visibility in order to mitigate the risks which it entails, these risks cannot be eliminated entirely. While visibility might be minimized, some level of visibility is also absolutely essential for cruising to work. When they cruise at beats, some men seek to make contact with other men who are 'strangers' – these efforts to make contact are a form of *public address*. While cruising men might deploy a kind of 'body language' which they hope will connect only with other cruising men, the scene which they construct and in which they participate is not a closed scene. The sexual vernacular of beat-cruising is not confined only to a select and known group, and while this form of address is not indefinitely accessible, neither is it predictably accessible. A wide range of people can potentially learn the particular language 'spoken' at beats, by recognizing the texts which circulate. The very characteristic of beat-cruising that constitutes a challenge to heteronormativity – its explicit publicity – is the characteristic that makes it both sexy and vulnerable.

So, I agree with Michael Bartos (1996) that while public health programmes are a form of governmentality in a Foucaultian sense, this does not make them 'bad'. Beat sex, as pleasurable as it is for beat users, is characterized also by the threat of violence, harassment, and the spread of HIV and other sexually transmitted diseases. If these characteristics eventually draw some form of response (as they surely will), what kind of response will it be? Will it be an increasingly violent repression of beat users by police? Will knowledge of the spatiality of alternative public sex cultures be produced only in efforts to curtail *any* form of public sex? Certainly, this has been the case in New York City, where some gay activists have assisted city authorities to identify and close sex-on-premises venues in order to prevent the spread of HIV (Colter, Hoffman et al. 1996). In response to such interventions, Warner (2000: 170) has argued that 'public sexual culture has to be a resource, not a scapegoat'. Clearly, the harm minimization approach offers a more progressive model of institutional involvement. This is particularly so when outreach projects are staffed by volunteers who are active participants in the counterpublic sphere in question, as they were in Melbourne. Project workers did not simply discipline beat users with a safe sex agenda, they (and the PLGLC) also supported beat users in conflicts with police and homophobic individuals.

Drawing wider public attention to beats also has the potential to widen the range of sexualities which can be expressed in the wider public sphere and urban public space. Nancy Duncan (1996: 138) has argued that 'lesbian and gay practices which potentially denaturalize the sexuality of public places could be more effective if they were widely publicized' rather than 'individualistic, privatized action'.[10] Publicizing alternative sexualities, as I have argued, is not without risks, but this does not make such a project fundamentally flawed. In protecting the space through which a publicly accessible alternative sex culture can be sustained, claims on behalf of beat sex can act as first step in constructing a wider 'queer' political project. For Berlant and Warner (1998: 548) the radical aspiration of queer politics is to build:

> not just a safe zone for queer sex but the changed possibilities of identity, intelligibility, publics, culture, and sex that appear when the heterosexual couple is no longer the referent or the privileged example of sexual culture.

To some extent, claims on behalf of 'safe' or 'legal' beat use in Melbourne fall into the category of making a safe zone for queer sex which may not challenge the spatial dominance of heteronormativity. Nonetheless, as Berlant and Warner (1998: 563) argue, if queers 'could not concentrate a publicly accessible culture somewhere, we would always be outnumbered and overwhelmed'.

The difficulty that remains is that efforts to legitimate beat-cruising mobilize forms of public address which are foreign to it, and potentially threaten its conditions of existence. Cindy Patton has spoken of this tension in her foreword to Michael Brown's discussion of AIDS activism and citizenship in Vancouver:

> The absolute materiality of cruising grounds and their occupation – by queers – is neither official nor representable, either in the sense of media representation (all we can see is people being arrested for doing something that should have been done elsewhere) or the sense of participatory democracy. The cruising body ceases to be a cruising body once it leaves Stanley Park [a cruising space in Vancouver]. It might be a gay man – even a gay MP! – who describes his cruising experiences. But in the moment of speaking for a class dislocated from its sole space of meaning, in the moment of democratically representing it as a class, this body ceases to be a *cruising* body (Patton in Brown 1997: xix–xx).

To put this in the terms developed in Chapter 2, we see here the tensions that arise when one simultaneously attempts to establish space for a form of public address through *tactical* appropriations of place and to sustain that form of public address by making *strategic* claims for its legitimacy through debates which engage with wider publics.

Of course, the attempts to legitimate beat sex through political channels that I have discussed in this chapter exist alongside a range of other strategies and tactics which are mobilized to establish space for queer possibilities. Beat-cruising exists in a complex relationship with forms of cruising which have been secured through money power at other commercial venues such as bars, adult cinemas and shops, and saunas. In recent years, the internet has emerged as an important new site for cruising, with over half of the 2,000 men surveyed in the Melbourne *Gay Community Periodic Survey* reporting that they occasionally or frequently used the internet to look for sexual partners (Hull, Van Deven et al. 2004). Each of these places sustains its own particular dynamics, and each is differently accessible for different men – which is to say that they all offer their own opportunities and constraints. Beats remain an important alternative to commercial sex-on-premises venues, which are typically located in expensive inner-city locations that may be inaccessible for many men. Those who make the trip from the country or the suburbs may find that high prices rule out many venues, while others may privilege the young and glamorous at the expense of other possible gay identities (Dowsett 1996; Hodge 1995). As one beats worker argued, there is a 'gay hierarchy of glamour and acceptability' to which beats provide an alternative. Support for beats in the wider public sphere, even when it relies on a form of governance through the production and use of spatial knowledge, might help to procure publicly accessible spaces for homosexual sex without the need to rely on gender, race and money power.

Chapter Five

Making a Name: Writing Graffiti in Sydney

Graffiti is everywhere. It is in the Bronx and in Belgrade, in Sydney and São Paolo. It is on walls and billboards, on train carriages and railway corridors, on buses and bus shelters, on telephone booths and toilet doors, on lampposts and post-boxes, on rubbish bins and removals trucks. This list could go on, and if the history of graffiti-writing is anything to go by, the list will never be finished. Graffiti mutates, and multiplies.

This chapter is concerned with this mutation and multiplication, as well as with the meaning, of graffiti. I am interested in exploring struggles over the conditions in which graffiti is practised as a form of public address. The spread of new forms of graffiti pioneered in the streets and subways of Philadelphia and New York City in the late 1960s and early 1970s to countless cities around the world provides a fascinating instance of the ways in which the production of urban forms of public address are bound up with transnational flows of culture, technology and people. In the context of such flows, city lives are simultaneously lived 'here' and 'there' (Smith 2001). As a consequence, forms of public address such as graffiti both work as a means to circulate texts and themselves circulate *as* texts, thereby facilitating their movement and reproduction across different social spaces. That is to say that graffiti, like other forms of public address, is not simply a vehicle for the circulation of the particular *content* of graffiti messages or images. When texts or images in the form of graffiti circulate, this is also the text-based circulation of graffiti itself as a *form* of public address. Of course, one could say the same of the novel or of film – the global movement of public texts in these forms has also been the global movement of these forms, which have themselves been adapted and mutated to serve a wide variety of purposes. Forms and flows are integrally related (Gaonkar and Povinelli 2003: 387).

The movement and reproduction of graffiti across social space, of course, is not a simple matter of replication. For one thing, as Gaonkar and Povinelli (2003: 392) have argued, 'a form can be said to move intelligibly (as opposed to merely physically) from one cultural space to another only in a state of translation'. And the movement and translation of graffiti from one space to another has been neither straightforward nor uncontested. Graffiti writers have had to identify and negotiate the physical opportunities for graffiti-writing in different contexts in order to make their mark(s), and this has inevitably brought them into confrontations with property relations and laws in a range of cities. These confrontations are absolutely critical in establishing the conditions of graffiti as a form of public address, and they have had distinct trajectories across different contexts. Graffiti writers' 'confiscation of the urban environment' (Stewart 1987) has been the occasion for costly graffiti wars which have been waged across numerous jurisdictions, on behalf of communities said to be under siege from 'anti-social behaviour' such as graffiti-writing. These wars have been waged with different weapons, and with different outcomes, in different places. A form-sensitive analysis of graffiti as public address must also then be a place-sensitive analysis of graffiti as public address. How does graffiti move? How is graffiti received and translated in different contexts? What kinds of struggles are fought over the conditions in which graffiti circulates, as both texts and forms?

In this chapter, I explore some of these issues through an historical account of the changing conditions of graffiti-writing in Sydney, Australia. The chapter begins with a brief account of the emergence of new forms of graffiti in cities on the east coast of the United States and how these forms came to be practised in Sydney. It then focuses on the various responses of urban authorities to the practice of graffiti in Sydney. As we shall see, both the practice of graffiti and its regulation relied on the public circulation of texts about graffiti and its meaning. Graffiti and anti-graffiti strategies have been publicly facilitated, discussed and debated through a huge range of media which are at some remove from the surfaces on which graffiti is actually written. Graffiti made appearances in Parliamentary debates, in mass media news reports and mass advertising campaigns, at police conferences and in criminological journals seeking 'solutions' to its proliferation, in small-circulation newsletters and glossy international magazines devoted to its ongoing production, in crime shows and music video clips trying to establish a credible connection with the 'street', in art galleries and design guides, and increasingly in expensive books and internet sites. The mediation which has made the circulation and translation of graffiti possible in Sydney has also made it governable. As such, graffiti writers and others involved in debates about graffiti have been particularly concerned with the nature of the 'counterpublic horizon' (see Chapter 2), as they have sought to expand or limit opportunities for (different kinds of) graffiti-writing to different audiences.

Writing Graffiti

The desire to circulate names – of politicians, of products, of celebrities, of brands, etc. – has long been a feature of public discourse and communication. The physical spaces of the city have of course been put to this purpose alongside other spaces of communication. Names have been circulated through soapbox speechmaking, market hawking, billboard and bus-shelter advertising, the use of spectacular lighting, and an ever-expanding range of urban practices, all of which are enmeshed in a complex web of property laws, planning codes and other technologies deployed to regulate the use of urban space. The writing of graffiti on urban surfaces, in some ways, bears a strong relation to these other forms of name-making through the city. The new styles of graffiti which emerged in Philadelphia and New York City towards the end of the 1960s by now have their own historians, drawn from the academy and from the ranks of graffiti writers themselves (see for example Austin 2001; Castleman 1982; Chalfant and Prigoff 1987; Cooper and Chalfant 1984; Powers 1999) – readers curious for a more detailed account now have plenty of sources to choose from. But it is necessary here to sketch out at least a brief account, to orient the more detailed analysis of graffiti's movement to Sydney which will follow.

The people who started writing their invented tag names on buses and trains in the late 1960s were not doing something wholly new. Graffiti has existed in some form or other for thousands of years, as the walls at the ruins of Pompeii attest. Sometimes this graffiti took the form of messages – of love, of hate, of injustice – while sometimes it took the form of personal or collective inscriptions which commemorated a visit to, or claimed, a particular territory. What was novel about the graffiti that emerged during the late 1960s and has continued to evolve since was not so much the writing of text on urban surfaces itself but its mode of application (with thick ink markers and aerosol cans), its frequency (with the same writer seeking to achieve widespread recognition through repeated and/or large-scale graffiti-writing) and its medium (with the use of moving objects, in particular subway trains but also trucks and buses, as media, as well as highly visible surfaces past which many people moved such as train stations and bus shelters). With growing numbers of writers competing for attention, styles evolved – their tags (invented, stylized signatures) became more elaborate, with the addition of arrows, characters, bubble-style lettering, colouring and 3-D effects. By the 1970s, single graffiti pieces could in some instances cover the entire external side-panel of a New York City subway car. The evolution of graffiti styles was facilitated through the emergence of localized graffiti-writing public spheres, which took shape as writers swapped sketches and photographs, hung out at 'writers' benches' (where groups of graffiti writers would gather on a bench at a particular subway stop to watch subway cars roll and view their graffiti), took exploratory subway rides to view the work of other writers, and formed collective 'crews' through which knowledge about graffiti-writing styles and techniques could be passed from writer to writer.

As the new styles of graffiti emerged and evolved, it was not long before they came to the attention of mainstream media and became the object of wider public discussion. At first, the representations of graffiti in such discussion were quite diverse. Some initial responses to New York subway graffiti tended to emphasize its novelty as a new form of folk art. Famously, graffiti writer TAKI 183 was featured in a 1971 article in the *New York Times* about graffiti writers, and with this article he and his peers gained considerable notoriety. And as graffiti on urban surfaces became the object of public discussion and debate, this further contributed to its circulation and growth:

> Shared public space (the streets) had already served as a broadcast medium for commercial advertising for more than a century by the time the new writing appeared on the urban scene. Following this lead, writers 'borrowed' shared public space to propagate their names. Now a second means had appeared: the commercial mass media could circulate a writer's name. Photographic, video, and film images in movies and television broadcasts as well as printed mentions of their work in newspapers and magazines flowed through the commercial public sphere. After TAKI's interview, dissemination in almost any form of commercial mass media offered a route to fame (Austin 2001: 49).

From its very beginnings, the circulation of graffiti texts in the new styles was mediatized, making them available to countless others beyond those who caught direct glimpses of them on the trains they used and the walls they passed. Indeed, Joe Austin (2001: 50) suggests that the mediatized representation of graffiti became an objective of writers, and thereby shaped their practice in important ways:

> Writers . . . deliberately sought to write their names in urban locations that they thought were likely to be photographed, videotaped or filmed. Many writers during [the 1970s] could boast that they had repeatedly appeared in the newspapers and on television, although these 'appearances' were usually made in the background of photos and camera shots intended for other purposes. Nonetheless, these appearances were proudly claimed and highly prized.

Here, then, physical sites in the city and various media combined as a 'space of appearances' (Arendt 1958) for graffiti writers.

But, as Austin has demonstrated, graffiti came to be portrayed in an overwhelmingly negative light by urban authorities and the mainstream media as the 1970s rolled on. While this shift in attitudes occurred simultaneously with the growth in graffiti, it also coincided with the fiscal crisis of the New York City Government. The eradication of graffiti was constructed as a key plank in the City's attempts to demonstrate its competence to govern:

> Graffiti became one of several symbols promoted as a stand-in for the sense that something fundamental had gone wrong, and its removal from the subways in the 1980s presented a visible task that could measure the tangible process of elite efforts to right the wrongs that elites themselves had created (Austin 2001: 5).

Mayor Koch's 'war' on graffiti mobilized special police intelligence gathering operations, razor wire, regular chemical washes for subway cars, guard dogs at subway lay-ups, and harsher penalties for graffiti writers and aerosol paint suppliers, alongside print and television advertisements with entertainment and sports stars trying to convince young people that writing graffiti was 'stupid' and 'uncool'.[1]

The problematization of graffiti-writing frequently invoked the property relation as a means to condemn the actions of graffiti writers. Graffiti was characterized as a form of unauthorized property damage, a cause and consequence of urban decay. The 'decay' that graffiti signalled was the decay of the very norms of respect for property ownership and urban authority. Trains could continue to run while covered in graffiti, which caused little functional damage to rolling stock. Indeed, functional damage would have been counterproductive – graffiti writers were just as concerned as the Metropolitan Transit Authority to ensure that rolling stock continued to roll. Rather, the 'problem' posed by graffiti was of a different nature. The appearance of graffiti came to be seen as evidence that authorities were not in control, that they were not in fact able to enforce their rights as the sole authors of the socio-spatial norms which applied on their property. In Michel de Certeau's (1984) terms, graffiti writers established a 'local authority' which imposed an alternative set of values on the surfaces upon which they wrote. The writers saw not 'property' (with all the social sanctions that concept entails) but a medium for circulating their names, artistic ambitions, and messages for each other and the wider public: 'The urban landscape became an unbounded billboard, a mass-mediated prestige economy pirated by the young' (Austin 2001: 47). The privacy rights of property owners (both 'public' and 'private') to determine the uses and appearance of the urban surfaces they possessed were actively ignored by graffiti writers. As such, graffiti can be 'considered a threat to the entire system of meanings by which such surfaces acquire value, integrity, and significance' (Stewart 1987: 168). As Susan Stewart went on to argue in her excellent (1987) essay:

> graffiti's confiscation of the urban environment, its relentless proclamation that what is surface is what is public, poses [a] threat to exchange, to business as usual . . .

The vilification and criminalization of graffiti-writing associated with various efforts to eradicate graffiti in New York City pushed writers to take further measures to protect their 'real' identities and put greater distance between this counterpublic and mainstream publics. But at the same time, some graffiti writers were finding other channels to reach a wider audience. As the 1983 documentary *Style Wars* shows, at the same time that graffiti writers were being targeted by urban authorities, they were being feted by art galleries and others in the wider music and art worlds. Writers like Dondi, Futura and Zephyr had their pieces hanging on gallery walls as well as circulating on the subway system.[2] Most significantly, the new graffiti styles gradually came to be associated with the nascent hip hop culture taking shape in the outer boroughs of the city. Some graffiti writers provided the artwork for flyers advertising the parties and clubs where new styles of breakdancing, DJing and rapping were evolving, others even mastered some of these cultural forms alongside their graffiti-writing, and the energies of those engaged in these various practices started to fuse. This fusion was absolutely critical for the circulation of graffiti as a form of public address beyond New York.

Graffiti Moves

If hip hop started out as a ghetto phenomenon in New York City, by the beginning of the 1980s it was beginning to reach a much wider, transnational audience. The mediatization of graffiti was essential to its circulation and eventual translation to other contexts. The year 1983 was a particularly big one for hip hop culture, with the gradual commodification of hip hop's cultural products enabling its exponents to tap into networks of commodity circulation with national and international reach. This was the year that Grand Master Flash and the Furious Five had mainstream chart success with their song 'The Message', which painted a vivid picture of life in the public housing projects of New York City. It was the year that the movie *Wild Style* and the documentary *Style Wars* first screened in the

United States – the first a movie which fictionalized the life of a New York graffiti writer with performances from many of the key writers, rappers and breakers in the New York hip hop scene, the second a documentary about the battles between New York urban authorities and graffiti writers over the subway system. Small quantities of these products travelled well beyond the United States. It was also the year that Malcolm McLaren's song 'Buffalo Gals' achieved massive sales not just in the United States, but in countries such as England, Australia and New Zealand. The video clip for this song had an impact on some people in these places far removed from the Bronx. Blaze, a participant in the early stages of the Sydney breaking and graffiti scenes, described hip hop's initial migration to Sydney in the following terms:

> Hip-Hop culture also migrated here, but not by boat or by plane. It came via television, cinema and radio, circa 1983/84. Like most other countries it came in a loosely held package. Strangely enough it manifested itself here via an Englishman's version of New York. Yes it was Malcolm McLaren's doing, more so it was the film clip to his 'Buffalo Gals' track. Although the song isn't all that, the visuals were. We heard the sounds of The Worlds Famous Supreme Team scratching, the Rock Steady Crew breakdancing and ... Dondi piecing up a Buffalo burner. Shit was too much at once (quoted in d'Souza and Iveson 1999: 58).[3]

Gradually, groups of Sydney young people also started breaking and listening to hip hop music. Crews of breakdancers united by a stretch of railway line that went from the city to the outer west began to congregate at Burwood Park in Sydney's inner west. DJ ASK, another early hip hopper, later recalled:

> it's not in the West any more, but it was back then, only because Hoyts [a major cinema chain] used to be across the road from Burwood park. They used to have *Breakdance* there, so everyone used to break in the park, after a while it turned into a big meeting place, and everything would happen there, especially Thursday nights everyone was there, and if you were outer west and you were in town, you'd come to Burwood. All the people I know always say Burwood's the home of hip-hop in Australia, even though it isn't now and there's no-one there really rapping or doing anything, that's where it all stemmed from for sure. All the people that used to break and rap were from those areas, as far as Bankstown and around the whole area. Back then, believe it or not, Burwood Westfield [a shopping mall] was the biggest one, and that was where you went to do your stuff and now it's just dwindled away because everything else got so big (quoted in d'Souza and Iveson 1999: 58).

Burwood was not only central because of its mall, but also because of its importance as a hub for Sydney's rail system, making it accessible to young people from all over Sydney. The existence of this rail network also provided the space in which graffiti-writing could flourish, allowing young people to experiment with reproducing the kind of graffiti they saw pictured in books like *Subway Art* – a book by photographers Martha Cooper and Henry Chalfant documenting hip hop graffiti in New York which was published in 1984 (Cooper and Chalfant 1984). The transmission of graffiti from New York to Sydney initiated a struggle over the conditions in which graffiti is written in Sydney that continues into the present. As we shall see, the transnationalism of graffiti as a practice was matched by the transnationalism of regulatory responses.

Graffiti began to attract regulatory attention as hip hop culture became more popular with some young people in Sydney. By 1986, the emergence of new forms of graffiti was of enough concern to NSW State Rail that it commissioned the Australian Institute of Criminology (AIC) to conduct research into the prevention of vandalism and graffiti on State Rail property. In contrast to contemporary state policies towards graffiti (to be considered shortly), the AIC Report is remarkable for its lack of direct engagement with hip hop culture. The report's introduction captures the emergent sense that graffiti was changing:

> both vandalism and graffiti are reported to have escalated in the Sydney area over the last few years and are now regarded as major problems by State Rail . . . Previously there has been a predominance of slogans and (often offensive) messages scratched or written in relatively easily removed ink, pencil, lipstick or crayon. This has given way to a predominance of 'tags' (stylized signatures), slogans and (still often offensive) messages written with felt tip markers and spray paint cans – which are much harder to remove. Less frequent, but equally hard to remove, are the large and colourful examples of New York-style art graffiti (stylized drawings of figures and the 'artists' names) done with spray cans of paint – usually on carriage exteriors, walls and buildings along the rail track (Wilson and Healy 1986: 2–3).

While the report acknowledged that new forms of graffiti had come to predominate, it associated this novelty with the new materials used in writing it (aerosol cans and markers) and the particular style that some graffitists seemed to have adopted. In other words, for the authors of the report, it was the *graffiti* that was new, not the *graffitist*. Indeed, the report argued that there had always been graffitists, and graffiti would not be going away 'despite its ephemeral nature and changing style and topics' (Wilson and Healy 1986: 33). As such, graffiti was framed as a sociological phenomenon that could be universalized across time and space. There was very little attempt to understand why new styles might have emerged, beyond some cursory remarks about the fact that they had been 'widely publicized and popularized amongst Australian adolescents through films, television and music video clips and books (eg *Subway Art*)' (Wilson and Healy 1986: 35).

The best deterrent to graffiti was thought to be quick removal, which (it was hoped) would frustrate the ambitions of writers for fame and recognition. The report therefore argued that trains and stations should be designed to prevent graffiti, and to make it easy to remove as quickly as possible. However, the authors noted a potential flaw in rapid removal strategies:

> The New York experience does, of course, suggest some limitations to this approach since the removal of graffiti can also provide a clean 'canvas' for more (Wilson and Healy 1986: 38).

Hence, it was also argued that young people should be involved in efforts to convert railway stations and carriages 'from public space to community space'. Based on the theories of criminologist Oscar Newman, the report argued that only by making State Rail property less anonymous could deviance be discouraged, by reducing opportunities for graffiti and vandalism and facilitating a sense of ownership by the local community. This would involve:

> deliberately encouraging groups in the local community – especially young people – to take an active role in using the local station for community purposes . . . or providing some input to its improvement . . . (Wilson and Healy 1986: 59).

This combination of law and order responses (designing out crime, increased surveillance, increased chance of being caught) and community development responses (providing more opportunities for legal public art) continues to characterize the approach of various state agencies towards graffiti in Sydney.

However, one aspect of the 1986 AIC report is more interesting for its *contrast* with contemporary graffiti policy. While trying to minimize the spread of vandalism and graffiti, the report also tried to put their impact in some perspective. Its authors argued that there was no statistical connection between graffiti and potential or actual violence on the State Rail network. The report also demonstrated that graffiti was much less costly than other forms of vandalism. In 1984, the estimated cost of graffiti removal from trains and train stations was around $635,000, while the estimated cost of other forms of vandalism (broken windows, slashed seats, smashed lights) was around $4.5 million (and while graffiti and other forms of vandalism might have been done by the same people, no such connection had been established). Nonetheless, the report acknowledged that the objective facts of graffiti's costs and relationship to violent crime were not matched by the perceptions of State Rail travellers, some of whom connected graffiti with a lack of personal safety. Of course, these perceptions were all-important in encouraging people to use public transport. With this in mind, the report recommended that State Rail actively attempt to disentangle violence and vandalism in their public statements about rail travel in the media:

> both staff and public must accept a certain level of rail vandalism and graffiti as inevitable – which they are. However, such acceptance is problematic – and probably unlikely – as long as vandalism and graffiti are seen as closely associated with violence (or even indicative of threatened or potential violence) on State Rail (Wilson and Healy 1986: 64).

Graffiti would never be totally eradicated, so State Rail was urged to make the distinction between graffiti and violence 'clearly, frequently and publicly' (Wilson and Healy 1986: 64). It was acknowledged that this project would be difficult, but in 1986 it still appeared possible – for while graffitists were generally regarded as delinquent, this did not necessarily imply that they were violent. The report reinforced this message with comments from young graffiti writers taken from a television show, emphasizing that they were not violent, but thought of themselves as artists doing graffiti for its own sake, rather than for some criminal purpose. While the report recognized that graffiti constituted a threat to the image of public space and public transport, it hoped to minimize that threat by mounting the public argument that it was only the *appearance* of these spaces and facilities that was slightly out of control – their safety and efficient operation were not threatened.

'Getting Tough' on Graffiti

The AIC Report's attempt to disentangle violent criminality from hip hop culture and graffiti appeared all but impossible by the late 1980s. Films such as *Colors* (1988), and the controversy over censorship of the government-owned youth radio network JJJ, which was banned from playing a song released by American rap crew N.W.A. called 'F*ck tha Police' (Wark 1990), firmly connected hip hop with gang violence. While no doubt graffiti would have been swept up in the generalized panic about hip hop, in 1988 the murder of a young woman in Sydney by a 'teenage railway graffiti gang' (*Sun Herald*, 11 September 1998: 3) helped to rocket graffiti writers to the front pages of Sydney's tabloid newspapers, and hence to politicians' law and order agendas. The *Daily Telegraph* ran a series of articles about graffitists, including a full-page spread reporting on 'the dark side of graffiti gangland' (*Daily Telegraph*, 13 September 1988: 4), and another article under the headline 'art of graffiti has lost its innocence' (*Daily Telegraph*, 17 September 1998: 4). The message, supplied by the recently established NSW Police Graffiti Taskforce (GTF), was clear – graffiti writers were violent criminals, and posed a serious threat to the safety of commuters and members of the public. The consternation shifted from the murder to graffiti itself. An editorial proclaimed:

> The murky world of the graffiti gangs is one of wasted young lives. It is also a world of vandalism and violence.
>
> It is a world authorities must eradicate.
>
> Thousands of youngsters – in hundreds of gangs – lead futile lives, leaving their marks only by the destructive gesture of spraying graffiti on trains and stations.
>
> They know the Sydney system better than the SRA.
>
> They know the sewerage system better than the Water Board.
>
> They know more about derelict railway property than all the accountants in the State Government . . .
>
> Their viciousness makes wiping out the gangs far more important than making sympathetic sounds about their sorry childhoods (*Daily Telegraph*, 13 September 1988: 10).

Some did make sympathetic sounds, arguing that the 'street kid' crime problem was worsening because of lack of funding for crisis accommodation in the inner city (*Daily Telegraph*, 20 September 1988: 4). Others, like criminologist Paul Wilson (one of the authors of the 1986 AIC Report), used another newspaper to express the view that it was 'a gross distortion of facts' to characterize graffiti writers as inherently violent (*Sydney Morning Herald*, 17 September 1988: 13). Even the public servant in charge of the NSW Office of Juvenile Justice concurred that 'It's outrageous that graffiti kids . . . have been labelled psychopaths' (*Sydney Morning Herald*, 17 September 1988: 13). Politicians nonetheless set about establishing new penalties for graffiti writers, and increased police resources devoted to transit crime. Police Minister Ted Pickering was reported to have said that 'special attention should be paid to graffitists', since 'recent experience had shown they were responsible for acts of violence' (*Daily Telegraph*, September 21 1988: 4). On announcing new penalties for graffiti offences, Minister for Transport Bruce Baird said:

> The fines are going to be quite substantial – the people who regard graffiti as an art form will find that it is no longer an art form and a matter which they are going to have to pay for significantly . . . The situation is out of control and we plan to address it (*Daily Telegraph*, 14 September 1988: 13).

The language of 'battle' and 'war' on graffiti, which the 1986 AIC Report had explicitly counselled against, became commonplace.

As that report had predicted, the war was a difficult one to win. Hip hop culture itself seemed to have taken hold among Australian youth. While by the late 1980s only faltering moves had been made by Sydney young people to produce their own hip hop music, American hip hop music was becoming more popular and widely available, and breaking and graffiti were also gaining in popularity. This growth in hip hop culture was sustained by independent record shops and alternative media, which secured the lines of communication from the United States to Australia, and facilitated the development of a local counterpublic sphere (Maxwell 2003). This counterpublic had its own publications like *Vapors*, which first appeared in Sydney in the late 1980s, and included photos of local, interstate and overseas graffiti. During the 1990s and early 2000s, graffiti styles and techniques also rapidly proliferated through the internet, as graffiti writers mobilized this medium to further circulate pictures of their work. Even if a piece of graffiti is 'buffed' from a site within hours of its completion, thus apparently reducing the likelihood of it being seen, photos of the work can nonetheless circulate widely in mediated form.

Of course, these counterpublic communication channels are accessible to 'outsiders' to the culture as well as 'insiders'. Graffiti magazines confiscated by members of the GTF were not casually tossed aside, but carefully studied (see *Weekend Australian*, 17–18 September: 4). Police, like graffiti writers, poured over hip hop publications from America and elsewhere, relying on the same channels of cultural flow that young people in Sydney had used to find out about graffiti in the first place. In order to better deal with the graffiti 'problem', they decided to learn more about 'murky world' of graffiti-writing and hip hop culture, and plugged themselves in to the emerging counterpublic sphere. The GTF had been established in 1988, some months before the panic about graffiti just described, following the recommendations of the 1986 AIC report concerning the need for better intelligence gathering. Its brief was to:

> patrol vulnerable areas and places where graffitist are known to congregate, identify graffitists by their tags and apprehend them, and respond to information from the public (Geason and Wilson 1990: 29).

As the GTF proudly proclaimed some years later, while some members of the public may not understand graffiti, the task force certainly did. They analysed writers' tags to 'work out who they are, work on patterns of where they operate, when they operate, and who they operate with' (*Sydney Morning Herald*, 18 March 1994: Metro, 1).

By 1990, the intelligence gathering activities of the GTF and other policing operations were clearly reflected in a new report published by the Australian Institute of Criminology: *Preventing Graffiti and Vandalism* (Geason and Wilson 1990). Based on GTF intelligence, the report presented more concrete statistical information about the extent of the graffiti problem than its 1986 counterpart:

> According to NSW police, there are at least 140 graffiti gangs or 'crews' with a membership of about 800 graffitists or 'writers' on police computers, but approximately 300 groups involving up to 3,500 youths from 12 to 18 years of age could be involved in the graffiti subculture in NSW alone (Geason and Wilson 1990: 7).

There was also a more sustained attempt to come to terms with the *culture* of hip hop graffitists, with discussion of 'crews' and 'writers'. Indeed, as well as presenting a section about 'the offenders', the report also featured a section on 'the culture'. The contrast with the 1986 AIC report is striking. Readers are informed that:

> Graffitists have a pecking order, and to qualify as a serious practitioner (a 'writer') rather than an amateur (a 'toy'), a youth has to spray his initials ('tag') at least 1,000 times on trains. If the train 'runs' with the tag still on it, this gives the writer more recognition among his peers, and one of a writer's great triumphs is to be photographed beside his handiwork before the railway maintenance workers clean it off (Geason and Wilson 1990: 9).[4]

There is an extended list of slang terms used by graffiti writers, which had been adopted by local graffiti crews from American sources along with 'the philosophical stance, designs and operational tips' (Geason and Wilson 1990: 10). To some extent, the discussion of a clandestine graffiti 'subculture' reflected the moral panic about hip hop culture that pervaded in Sydney at the time:

> The graffiti subculture seems to be highly developed, with secret signals, symbols and handshakes and cult books. There are reports of information exchanges, crew summit meetings, extensive interstate and even international connections, and a monthly periodical called *Hip Hop* (Geason and Wilson 1990: 10).

Nonetheless, the authors also attempted to point to the problems inherent in the outright vilification (or glorification) of writers:

> The current polarization of the debate about graffitists into dangerous criminals versus high-spirited kids with a love of public art may be preventing a

serious examination of graffiti as a system of extensive alienation, hostility and social malaise on the part of growing numbers of youngsters (Geason and Wilson 1990: 11–12).

As well as picking up tips on hip hop culture through reading American publications influential in the local scene, criminologists and police consulted transport and law enforcement agencies from cities like New York and London. On the basis of overseas and local experience, the authors concluded that

preventing or minimizing graffiti and vandalism seems to depend on the right formula, or package, of measures – police or railway police presence, electronic surveillance, quick and effective clean-ups, education campaigns, restrictions on the weapons or tools used, and programs and activities that prove more attractive to young people than 'bombing' trains or hanging around railway stations (Geason and Wilson 1990: 27).

A number of these measures had already been put in place by State Rail following the 1986 recommendations. The new carriages that had been introduced into the network since the late 1980s included graffiti-resistant materials, making them easier to clean (particularly of tags or pieces on the carriage exterior). New security and surveillance measures (such as barbed wire fencing, lighting and closed-circuit television surveillance cameras) were introduced in rail yards and stations in the attempt to prevent graffitists painting trains at night. More resources were devoted to early removal, and the GTF had established a presence on the rail network. The head of the NSW Police Transit Crimes Unit, in which the GTF was based, conducted a five-week research trip to the United States to study police methods concerning street gangs and transit crime (Hickman 1991).

As predicted in 1986, rail authorities and police have not 'won the war' against graffiti on State Rail property, although they have certainly succeeded in reducing its visible appearance and made rolling stock and railway corridors a more challenging target. This 'success', however, had unforeseen consequences. As one City Rail spokesperson acknowledged, when the train network became a more difficult target in the early 1990s, graffiti writers began to find other canvases:

Whilst there is still graffiti on our trains in 1994, it seems to be moving more to walls, shops and things like that. Hopefully, that means that they are moving away from us because we are making it harder for them, and whilst that pleases us, it doesn't please the community (*Sydney Morning Herald*, 18 March 1994: Metro, 1).

As the tactical opportunities for graffiti on the public transport network were restricted, graffitists began to make use of surfaces in the city more generally.

Indeed, graffiti writers have demonstrated a remarkable capacity to 'map' new opportunities for 'getting up'. Graffiti-writing is a cartographic practice as much as anything else, and these cartographic skills extend to keeping abreast of new regulatory responses and technologies which constantly transform the terrain in which writers work. So, by the 1990s, as incidences of graffiti-writing beyond the rail network became widespread in Sydney, graffiti became a 'problem' for a much wider variety of agencies and actors whose property was put to work by graffiti writers – local governments, public utility companies such as Australia Post and the Roads and Traffic Authority, schools, shopkeepers, outdoor advertising agencies, householders with fences or walls in high-profile locations, and many more. And as the surfaces upon which graffiti is written have multiplied, so have its forms of application. For example, one graffiti removal operator recently noted: 'now that spray cans are harder to come by [you must now be over 18 to purchase them], glass scratching is becoming popular for taggers' (*Daily Telegraph*, 26 April 2006: 42). Rapid removal has also been matched with rapid application methods such as stickers and stencils, which enable graffiti to writers to get up in situations where lingering to complete a more complex piece would most likely result in capture.

In the face of this multiplication, the repressive measures which seemed to have had some success on the rail network were further developed in order to deal with graffiti more generally. The NSW State Government did what it could to develop measures which could apply to graffiti-writing across the city. Graffiti figured as a significant issue in both the 1995 and 1999 NSW election campaigns. In November 1994, the *Daily Telegraph Mirror* ran a series of articles about youth crime under the banner 'City of Fear' (beginning on 21 November 1994: 4–5, 11). Shortly after these articles appeared, then opposition leader Bob Carr launched Labor's Law and Order policy, with its now infamous pronouncement that:

> We simply can't allow roaming groups or gangs of youths, their baseball caps turned back-to-front, to stop citizens walking the streets, shopping or using public transport (Australian Labor Party (NSW) 1994).[5]

Not to be outdone, Premier John Fahey rapidly introduced two new pieces of legislation to Parliament to position himself for the approaching election – the *Children (Parental Responsibility) Bill 1994* and the *Summary Offences and Other Legislation (Graffiti) Amendment Bill 1994*. The first was intended to give police the power to remove young people under the age of 16 from a public space if they thought that it would 'reduce the likeli-hood of a crime being committed'. The second introduced new penalties of up to 12 months' imprisonment for graffiti, and up to 6 months' impris-onment for 'possession of a spray can with intent'. The Labor Opposition, after some dissent within its ranks about law and order policy, eventually weakened the bills, introducing a trial period for the new police powers and halving the graffiti penalties, although it continued to support them in prin-ciple (O'Sullivan 1995). The irony of this whole situation was that the 'City of Fear' series of articles had been inspired by a leaked report into gang violence commissioned by the NSW Police. The report, later released pub-licly, argued that graffiti crews were significantly different from organized criminal gangs, and stressed once again that there was no inherent con-nection between graffiti writers and violence (parts of the report are repro-duced in Healey 1996).

In 1999, a similar auction of law and order policies characterized the election. Opposition Leader Kerry Chikarovski opened the bidding, fol-lowing a 'graffiti attack' on the War Memorial in Hyde Park in February 1999. She promised tougher penalties for graffiti writers, and the re-establishment of the police Graffiti Taskforce that had been disbanded by Labor.[6] Premier Carr responded with promises of more funding for a 'Beat Graffiti' scheme, and with a proposal to make young offenders serving Community Service Orders (CSOs) available to local government author-ities for the removal of graffiti (*Sydney Morning Herald*, 11 February 1999: 8). Chikarovski bid again. Launching her Neighbourhood Policing Policy (while standing in front of a wall adorned with graffiti tags), she proposed legislation which would allow magistrates the discretion of six-month jail sentences for serious graffiti offenders, regardless of whether or not they were before the magistrate for their first offence (*Sydney Morning Herald*, 10 March 1999). Not to be outdone by Carr's promise of teams of young offenders cleaning up graffiti, she promised to make them wear fluorescent vests branded with the words 'Community Service' while they scrubbed (*Sydney Morning Herald*, 22 March 1999: 6). Chikarovski lost the election. In March 2000, after the relevant legislation had been passed, the Juvenile Justice and Local Government Ministers announced a plan to make 66,000 hours of free CSO labour available to local governments for graffiti removal, posing for the assembled media in front of a team of young offend-ers cleaning graffiti on a wall in Seven Hills (*Daily Telegraph*, 17 March 2000: 7). Within a week, the wall boasted a new set of tags, but a spokes-person for the Juvenile Justice Minister said that the government would not be deterred from its 'war against graffiti vandals': 'If we go back every time they paint, eventually we will win and they won't come back' (*Daily Telegraph*, 26 March 2000: 11). War continued to be an important part of the language of graffiti policy.

And the war is far from over. The government claims there have been successes, but the fight to 'eliminate the scourge of graffiti from our communities' goes on (see *NSW Legislative Council Hansard*, 10 May 2006). Like all wars, the war on graffiti is costly. In 2004–5, the combined spending of Rail Corp and the City of Sydney alone on graffiti prevention and removal was over AUS$4million – and this does not take into account the millions more spent by other local governments, public utilities and private property holders (see *Sydney Morning Herald*, 27 December 2005: 3, and 21 January 2006: 13). In that year, Rail Corp NSW (2005) estimated that it removed 121,000 tags from its trains. Opposition transport spokesperson Barry O'Farrell described Sydney's train carriages as 'mobile ghettoes' (*Sydney Morning Herald*, 27 December 2005: 3), in a comment which appeared to be part of yet another campaign by the Conservative opposition to suggest that Labor was 'soft' on graffiti and other forms of anti-social behaviour.[7] New Labor Premier Morris Iemma responded with the formation of a State Government multi-agency 'Anti-graffiti action team' to coordinate anti-graffiti measures, and Rail Corp announced that it would kick in a further AUS$500,000 to assist NSW Police with 'Operation Chalk', its latest anti-graffiti operation (see *NSW Legislative Council Hansard*, 9 May 2006). The government also announced plans to further restrict the accessibility of aerosol paint, by legislating to require retailers to place spray paint either in locked cabinets or behind a counter (see *NSW Legislative Council Hansard*, 9 May 2006). NSW graffiti politics is nothing if not predictable.

Graffiti or 'Aerosol Art'?

The policies designed to 'get tough' on graffiti described above have coexisted since at least the beginning of the 1990s with the institutionalization of attempts to develop less repressive graffiti 'solutions'. The 1990 AIC report, after listing the package of responses conventionally adopted in overseas cities to address the graffiti problem, also went on to note that:

> A less traditional response – and one that seems to be successful with some graffitists at least – is mounting programs which take graffitists' artistic aspirations seriously and offering them a legal outlet for their art (Geason and Wilson 1990: 27).

The report noted that following the panic about graffiti in September 1988, 'graffitists and their sympathizers fought back, and stories portraying graffitists as, if not quite the kid next door, at least frustrated artists, began to appear in the media' (Geason and Wilson 1990: 8). Calls were made for the recognition of hip hop-style graffiti as 'aerosol art', and youth advocates in particular fought for the provision of legal opportunities for writers to develop this art.

The construction of claims on behalf of legal graffiti opportunities was helped by the confluence of a number of factors. First, as some early graffiti writers reached legal adulthood, the penalties for getting caught became much more serious, with large fines and potential prison sentences. As a consequence, some simply stopped writing, while others whose reputation was already established began to seek privately commissioned opportunities for legal graffiti (see for example *Bulletin*, 5 December 1990: 70–3). Second, during the mid-1980s workers in some community-based youth centres (like those in Bondi and Marrickville – see Geason and Wilson 1990: 27 and *Sydney Morning Herald*, 15 March 1996: Metro, 5) had begun to offer opportunities for young people to write graffiti. These services supported hip hop as a powerful form of individual and collective expression, and in some cases formed connections with older graffiti writers. Third, these services usually had close relationships with local government authorities, who had recently found themselves having to deal with graffiti on properties within their boundaries. These local government authorities had less scope to introduce the conventional responses of their State counterparts, such as tougher penalties and more policing, and as a result they were open to alternative approaches.

Marrickville Youth Resource Centre's (MYRC) Contemporary Urban Art Project is an example of the kind of alternative, less traditional approach to the graffiti 'problem' that emerged in Sydney (see Strong 1997). MYRC first become involved in providing legal opportunities for young people to develop their graffiti-writing skills as early as 1984. In 1996, a proposal for funding was developed by MYRC and Marrickville Council and Cellblock Youth Health Centre to consolidate their efforts to provide legal graffiti opportunities for young people 'at risk' of entering the juvenile justice system. On receiving funding for a six-month project from the Department of Juvenile Justice, an established professional graffiti writer was employed to run weekly sessions for participants.[8] As well as providing paint and space to write, the sessions focused on the history of graffiti and hip hop culture, sketching, and technique. The six-month project formally concluded with an exhibition in a Council-owned art gallery (see *Sydney Morning Herald*, 15 March 1996: Metro, 5), although more materials were bought and wages paid after this initial period through the revenue raised from some commercially commissioned legal work. MYRC's involvement in this project reflects its broader support for hip hop culture. The Centre also runs breakdancing sessions (Strong 1997), and it has acted as a venue during the *Urban Xpressions* hip hop festivals organized by Sydney hip hop artists (*Sydney Morning Herald*, 20 March 1998: Metro, 5). MYRC's embrace of hip hop culture continues to receive support from both local government and the local hip hop scene some ten years after the first project.

Not all legal graffiti opportunities have had the same level of support from local government authorities. Throughout the early 1990s, for example, Tony Spanos made the car park of his meatworks in Alexandria available for graffiti writers – a space that came to be known as the Graffiti Hall of Fame. Spanos also helped to facilitate legal works for some of the writers he met through the Hall of Fame (see for example *Sydney Morning Herald*, 9 July 1992: 14). Spanos developed a strong local profile, and was initially supported by the South Sydney Council in his efforts to provide a legal space for graffiti-writing. But towards the end of the 1990s, Spanos and the Graffiti Hall of Fame fell out of favour with the Council. Not only did he allow graffiti, he also allowed the space to be used for all-night dance parties. When the area around his meatworks was re-zoned from industrial to residential, these parties became the subject of a series of court cases in the Land and Environment Court (*Sydney Morning Herald*, 19 November 1999: 9), which resulted in the eventual closure of the Graffiti Hall of Fame to graffiti writers and dancers alike. Spanos, like Strong (1997), emphasized the importance of space being made available for young people through such projects and initiatives:

> A car park is all I have given these kids and they created their own energy. The Government has billions of dollars and all they needed was a car park in a meatworks (*Sydney Morning Herald*, 19 November 1999: 9)?

At both MYRC and the Graffiti Hall of Fame, a 'proper place' for graffiti has been created through the use of individual property rights. In both cases, this has required difficult negotiations with government planners. In Spanos' case, the property rights usually invoked to argue *against* graffiti have been restricted when an attempt was made to use them to *facilitate* graffiti. Graffiti was assumed to have a negative impact on neighbouring properties, so that even when a property owner tries to make a place for graffiti it is still defined as 'out of place' in the neighbourhood in which that property is located.[9]

Locally based legal graffiti/aerosol art projects provided a base from which some youth advocates mobilized campaigns in support of hip hop-style graffiti at the larger scale of State politics. Between the 1995 and 1999 elections, Premier Bob Carr established the *Graffiti Solutions* project, in an attempt to coordinate the different graffiti strategies being undertaken at various levels of government. The project was launched in 1997 at the *Graffiti: Off the Walls* conference in Sydney.[10] A community panel was set up to advise the government on its handling of the graffiti issue, and it has overseen the funding of various trial projects at the local government level. Youth advocates (including Carole Strong and Matthew Peet from MYRC) already involved in local legal graffiti projects used the opportunity provided by the conference and the community panel to push for more 'inclusive' solutions to the graffiti problem. In particular, youth advocates emphasized the cultural dimension of hip hop-style graffiti, in attempts to address the invisibility of the cultural context in which young people write graffiti (Collins 1998; Peet 1998; Strong 1997). Advocates have been relatively successful in making a distinction between legal and illegal graffiti, and a number of local government pilot projects have established legal graffiti spaces as part of their response to graffiti (Curtiss 1998; Pinnington 1998). These projects, funded under the 'Beat Graffiti' grants scheme, are under review by the NSW State Government at the time of writing.

Legal graffiti projects providing 'aerosol art' opportunities are not always framed as alternatives to 'get tough' policing strategies – indeed, sometimes they are framed as making a valuable contribution to such strategies. In his paper to the *Graffiti: Off the Wall* conference, Newcastle youth arts worker and police officer Andrew Collins (1997) categorized hip hop graffiti writers as either 'traditionalists' or 'modernists'. Traditionalists, he said, were totally opposed to the idea of legal graffiti, preferring to keep it street-based. The modernists were apparently more interested in developing their art form, and more likely to be open to involvement in legal graffiti projects. In Newcastle, his local community arts project had successfully worked to 'modify crew behaviour to reorient the focus of an individual culture from traditional to modern' (Collins 1997). Project workers had done this by helping in the arrest of the traditionalists, and promoting opportunities for modernists, so that they rapidly 'infiltrated' the ranks 'until an entirely modernist hierarchy existed'. He told conference delegates that 'strategies run through arts bodies or youth organizations provide an ideal intelligence source for law enforcement bodies' (Collins 1997). In this model, promoted for adoption in Sydney and elsewhere (Collins 1998), police strategies have shifted from an understanding of hip hop culture (as developed by the GTF) to a conscious attempt to infiltrate and shape the culture.

Attempts to legitimize any form of hip hop-style graffiti as 'aerosol art' have met with predictable opposition. For example, Leo Schofield, former director of both the annual Sydney Festival and the Olympic Arts Festival in 2000, regularly used his weekly column in the *Sydney Morning Herald* to lambaste graffiti writers and their supporters. His hyperbole has sometimes bordered on the kind of gangsterism he associates with 'spray can vandals':

> While defacement and desecration continue apace, there are those in the 'arts community' who are still trying to legitimize graffiti as an art form. This week

I noticed another reference to 'Aerosol Art'. How pleasing it would be to stumble upon a group of these 'artists' defacing public property. One could then perhaps exercise one's own artistic inclinations by covering the sneaky little s---s from head-to-toe with some of their own paint. Better still, given their fondness for the aerosol can, hit 'em with a few bursts of capsicum spray or mace (*Sydney Morning Herald*, 14 March 1997: 32).

Not surprisingly, the Sydney 2000 Olympic Games provided few opportunities for graffiti writers to showcase their talents. Indeed, the Olympic Games were used by graffiti opponents to justify an expansion in more conventional approaches to the graffiti problem – just as they had been used in Los Angeles in 1984 (Stewart 1987).[11] A journalist for the *Daily Telegraph* (17 January 2000: 8) described the train line from Central Station to the Olympic site as 'Sydney's eyesore express', but was assured by the Transport Minister that plans were in place to deal with the issue. The Minister later told Parliament that 320 cleaners were engaged in a two-month period before the games to clean the rail corridor (focusing on the Olympic corridor), and 60 were employed throughout the Olympics to stay on top of the problem. Sydney City Council also contracted a firm of 'graffiti busters' to ensure that all graffiti in the CBD was erased within 24 hours during the Olympics – the same 'crack anti-graffiti flying squad' that had been on '24 hours alert to clean up graffiti as soon as it appeared' during a visit to Sydney in early 2000 by International Olympic Committee President Juan Antonio Samaranch (*Daily Telegraph*, 22 January 2000: 23; and 1 February 2000: 12).

Graffiti and the Market

Debates between those who would have the state 'get tough' on graffiti and those who would have the state provide opportunities for 'aerosol art' are in some ways reflected in the marketplace. The fact that graffiti continues to be outlawed lends the form a certain cultural credibility that is attractive for those seeking access to the 'youth market'. Clothing companies, record companies, food and drink providers and others – local, national and global – now regularly make use of graffiti styles to attach a 'street' aesthetic to their products. Shop-fronts, packaging, advertising, and commodities themselves are adorned with graffiti-style text and imagery. Graffiti techniques are deployed in street-based marketing campaigns – in one controversial instance, a national telecommunications provider in which the Commonwealth Government is the majority shareholder even deployed a stencil-graffiti advertising campaign on footpaths all over Sydney (*Sydney Morning Herald*, 30 October 2004). And graffiti writers (and legal graffiti programmes) in Sydney also provide a lucrative market for commercial providers of ink markers, aerosol paints and other 'tools of the trade' – some of which are now explicitly designed and marketed with graffiti writers in mind. This marriage between graffiti writers and commerce in Sydney, as in other cities, provides a further set of opportunities for the graffiti writers. In this way, the commodification of graffiti has further contributed to its circulation as a form of public address.

As such, the market for graffiti and graffiti-related products has not escaped governmental interest and regulation. Most obviously, this has been pursued through efforts to reduce the sale of spray paint to graffiti writers which were detailed earlier. Further to these efforts, in May 2006 NSW Attorney General Bob Debus called on the Commonwealth Government to crack down on the sale of spray cans via the internet:

> those who do graffiti are increasingly buying their spray cans over the Internet. Apparently the most desirable paint comes from Germany. We need to take actions to ensure that the free market of the Internet does not undermine our best efforts, and only the Federal Government can do something about that (*NSW Legislative Assembly Hansard*, 9 May 2006).

In seeking to enlist the Commonwealth's help, Debus had reason to think he might receive a sympathetic hearing. In February 2006, Australia became the first and only nation to ban a graffiti-related computer game – Marc Ecko's *Getting Up: Contents Under Pressure* – on the grounds that it incited criminal activity. At Commonwealth Attorney-General Phillip Ruddock's request, the Office of Film and Literature Classification's Review Board denied classification for the game, thus making it illegal for it to be sold, demonstrated, hired or imported (*Sydney Morning Herald*, 17 February 2006: 3). This judgement was justified on the grounds that the game facilitated the 'transfer of the game-world knowledge of graffiti to the real world', and that the 'crime of graffiti' was thereby 'glamorized and normalized' (Classification Review Board 2006: 12, 13).

Of course, the graffiti prevention and removal business is now also a lucrative multi-million-dollar industry in Sydney. As public and private sector agencies contract out their graffiti removal to private companies, a mini-industry has emerged. The 2006 Sydney phone directory listed 35 companies specializing in graffiti removal. Miguel d'Souza, who wrote hip hop column *Funky Wisdom* in the Sydney Street paper *3D World* (27 April

1998), had a novel suggestion for graffiti writers as this industry began to take shape:

> Of course if writers here had any sense of irony, they'd go into the graffiti-removal business and bid for government tenders. After all, they have an idea of how the stuff goes on, surely there's some value these days, when work is hard to find, to apply some experience and knowledge gained; and once you've removed it, someone can go about the business of putting something beautiful up there to replace it. Isn't capitalism just the best?

'Keeping It Real'

Not surprisingly, the increase in opportunities for legal and commercial graffiti has been a subject of sustained debate among graffiti writers. Are the graffiti opportunities created by aerosol art advocates in youth centres and private companies 'real' graffiti? Are legal walls and corporate adver-tisements 'proper places' for graffiti? And do graffiti writers risk exposing themselves and/or their culture to repression and exploitation through par-ticipation in legal graffiti projects – particularly when some have been explicitly linked to police intelligence-gathering efforts? The comparative merits of illegal and legal graffiti have been debated by graffiti writers and others with a sympathetic orientation towards hip hop culture through the spaces and media of the counterpublic sphere. Clearly, both legal and illegal graffiti take a variety of forms. But despite this variety, at least three sig-nificant points of difference between legal and illegal graffiti have been articulated in the course of these debates. First, illegality is valued as a virtue by at least some graffiti writers for its own sake. Ian Maxwell's study of Sydney hip hop reports that tales of brushes with the law, and stories of escape from (or indeed capture by) the GTF, constitute a kind of cultural capital within some elements of Sydney's hip hop community – they are a marker of commitment to graffiti not just as a style, but as a *lifestyle* (Maxwell 1997). Nancy MacDonald (2001) finds a similar dynamic at work in her sociology of graffiti writers in London. The *practice* of graffiti, then, is just as important as the *product*.

Second, the style of the graffiti itself often tends to be different in legal projects. Certainly, in spaces like the Graffiti Hall of Fame and MYRC, graffiti writers work for themselves and each other, not for the wider public. But on legal walls and in commercially commissioned work, writers are typ-ically required to develop a style to suit public tastes, to make the graffiti legible to those not familiar with hip hop style. Further, claims for the recognition of hip hop-style graffiti as art generally have had much more success in gaining acceptance for colourful pieces rather than tags. In an effort to legitimate the provision of space for legal graffiti, youth advocates have sometimes characterized tagging as a 'stage' that a writer goes through on their way to writing murals or pieces – one which might be circumvented if they can progress straight to pieces with the help of community arts projects (see for example Peet 1998). This fits prevailing public attitudes, first identified by the AIC in 1986 (Wilson and Healy 1986: 34–5), which seem less troubled by pieces than by tags. No doubt such attitudes prevail because police and urban designers continue to assert that tags are associated with a violent, dangerous criminality – in a perspective more informed by *West Side Story* than any serious analysis, tags are supposed to signify a gang 'marking their turf' (see for example *Sydney Morning Herald*, 12 January 1999: 28; see also 18 March 1994: Metro, 1) – notwithstanding evidence to the contrary in the criminological research discussed earlier. However, some graffiti writers continue to value (artistically and culturally) the practice of tagging as highly as piecing. Snarl, for example, in the ABC hip hop documentary *Basic Equipment* (1998), said:

> People come up to me and say 'I like this stuff [referring to pieces] but I don't like the scribble.' But if you're going to like this, you've got to look at tagging too, and take everything, because it's all part of one thing. You can't take any element and say 'I like this but not that,' you know?

Third, legal graffiti projects reinscribe a respect for private property relations, and the consequent control by owners over the appearance of public spaces in which they are located. This applies as much to the private property rights of state agencies like the Rail Access Corporation as it does to the private property rights of individuals and privately owned enterprises. As noted earlier, by tactically challenging the prerogative of private owners to control the appearance of their walls, graffiti draws attention to the very fact of this control. Of course, writers do not do this in the name of a universal public or common good. Indeed, graffiti writers have often expressed their desire to seek recognition from others within the hip hop community itself, rather than from the general public. As one writer put it bluntly in *Basic Equipment* (1998), 'if mainstream society doesn't like it, they can go and get fucked'.[12] Similarly, another writer more politely informed a journalist from the *Sydney Morning Herald* that 'The only opinions that matter are those of other writers' (18 March 1994: Metro, 1). To negotiate a legal opportunity for a piece is to bring 'mainstream society' firmly into the equation. It is to accept that private owners have the right to determine the face their premises present to public space, thus helping to constitute public space itself. A legal mural might have some effect in making young people's culture more visible in public space, but often this is on someone else's terms. As a Canadian academic specializing in Crime Prevention Through Environmental Design (CPTED) who visited Sydney in 1999 said, legal murals 'can positively contribute to an interesting sense of *controlled disorder* that makes a place interesting' (*Sydney Morning Herald*, 12 January 1999: 28, my emphasis).

The different positions articled in these debates have material consequences for the different forms of graffiti that are practised in Sydney. In particular, the attitudes and exploits of writers who continue to value illegal graffiti over legal graffiti often spill over from the hip hop counterpublic sphere into the wider public sphere. This is a source of considerable tension, as different representations of hip hop culture compete for attention. For example, following two incidents involving the tagging and piecing of public buses in depots, the *Daily Telegraph* did some 'investigative journalism' about graffitists. In an article which combines panic about young people's use of the internet with moral indignation about graffiti, the article's headline alarmingly reported that 'spray can gangs brag on the net'. Accompanied by colour pictures taken from the offending websites, the article is contemptuous of any efforts to consider graffiti as art rather than vandalism:

> Graffiti vandals are bragging of their exploits on the Internet where some of Sydney's most notorious young defacers are posting images of their work.
>   Photographs of defaced trains, paint-splashed walls and lists of the most durable paints available are posted on Websites identified by *The Daily Telegraph* yesterday. One Website even includes a map of the City Rail network to make it easier for vandals to target trains, stations and neighbouring buildings . . .
>   One site describes vandals as 'writers' and justifies their 'art' as 'modern day hieroglyphics' (*Daily Telegraph*, 5 January 2000: 5).

The perspectives of those who value illegal graffiti threaten to undermine efforts to gain recognition for pieces as a legitimate form of art by other youth advocates.

Hip hop-style graffiti now exists at the intersection of legal and illegal work, in a creative tension which is sometimes uncomfortable, sometimes productive. As Phibs, a former Sydney writer now based in Melbourne, has said of attempts to consider graffiti as 'art':

> When you're doing it, when you see an amazing piece, you know it's taken time, effort and talent. You know it's art and you respect it and respect the artist. But it's not in a gallery, not being bought and sold by the capitalists. We don't do it to get paid – we do it for ourselves. It's our high!!! We don't represent, we signify!!! . . . Untrained youth usually do it, not recognized as artists, but more as urban pests and mess-makers, and the fact is that it's a crime against property and society . . . That is an important part of what it is – it's energy and motivation, it's spontaneity and immediacy. It's here today, gone tomorrow – maybe (*Head Shots*, no4, May 1997: n.p.)!

For writers like Phibs, 'real' graffiti is both artistic *and* outlawed. His approach to legal opportunities reflects this dialectic:

I will always enjoy the opportunity to paint a legal wall that I can take my time on or get paid for, but for real graf to keep growing strong it's got to stay underground – it's got to retain its illegal, criminal edge (*Head Shots*, no4, May 1997: n.p.).

Phibs' pieces and stickers are all over walls and other surfaces in Sydney and elsewhere, some legal and others not. His work has recently been featured in the 'outdoor gallery' space in May Lane (see Box: May St, St Peters), and in a billboard and bus shelter advertising campaign for a new flavour of Absolut Vodka.

May St., St Peters

May St., in Sydney's inner western suburb of St. Peters, is host to a remarkable diversity of graffiti styles and practices, as well as a variety of graffiti responses. Tags and throw-ups adorn blank walls, fire-escape doors, garage roller doors, Roads and Traffic Authority signal boxes, and windows of vacant buildings – some are well practised, others are not, and presumably none are there with permission (see Figures 5.1 and 5.2). The owners of vacant buildings and the

Figure 5.1   Garage door tags, May St. (photo: Kurt Iveson)

Continued

Figure 5.2   Throw-up, May St. (photo: Kurt Iveson)

Figure 5.3   Graffiti Clean, graffiti-removal contractor, May St. (photo: Kurt Iveson)

occupants of some semi-industrial premises seem content to let this graffiti stay, while the owners of a recently completed apartment complex engage private contractors to remove graffiti as soon as it appears (Figure 5.3). Others have engaged graffiti writers to do more artful pieces, in the hope of reducing further tagging (Figure 5.4). In May Lane, the owners of a print shop have established their walls as a gallery of contemporary graffiti styles, engaging well-known graffiti

Figure 5.4   Legal Commission, May St. (photo: Kurt Iveson)

Figure 5.5   Graphic Art Mount wall, May Lane (photo: Kurt Iveson)

writers to do new pieces every month. These pieces are often less traditional, combining hip hop letter styles with stencil work, characters and more abstract themes (Figures 5.5 and 5.6). The owners of the May Lane print shop have a website documenting these pieces

Continued

Figure 5.6   Graphic Art Mount wall, May Lane (photo: Kurt Iveson)

Figure 5.7   Wall opposite Graphic Art Mount, May Lane (photo: Kurt Iveson)

(http://www.mays.org.au/), and in May 2006 they staged a retrospective exhibition with a glossy printed catalogue. The rest of the Lane has emerged as a kind of 'hall of fame', with other writers completing large-scale pieces, stencils, posters and tags on nearby surfaces (Figures 5.7, 5.8 and 5.9).

Figure 5.8 · Assorted graffiti, May Lane (photo: Kurt Iveson)

Figure 5.9   Assorted graffiti, May Lane (photo: Kurt Iveson)

> Of course, it is difficult to convey the colour of this street in black and white! These photos of May St (and more) can be found in full colour at: http://www.geosci.usyd.edu.au/about/people/staff/iveson.html.

Conclusion: The Graffiti-writing Counterpublic Sphere and the Audiences of Graffiti

In November 1998, one 18-year-old graffiti writer was caught by two security guards in the railway yards in Redfern, an inner city locality in Sydney. Before calling the police, they 'kicked him to the ground and beat him with batons, leaving him with cuts, bruises and welts', then proceeded to steal his camera, and pose for photographs with the writer pinned between them in front of a noticeboard on which had been written 'graffiti artist caught at Eveleigh Railyards' (*Sydney Morning Herald*, 5 May 1999: 4). Perhaps they were fans of Leo Schofield's newspaper column. Writing graffiti entails risks which are sometimes serious.

In this chapter, I have been concerned to understand the nature of struggles over the conditions in which graffiti is written – charting how the opportunities for graffiti-writing have been shaped and reshaped. With a variety of agencies and institutions taking an interest in the conditions of graffiti-writing in Sydney, the 1990s and early 2000s bore witness to a wide array of interventions calculated to shift the practices of graffiti writers. Significantly, while the resources devoted to waging war on graffiti reached spectacular heights, claims for the legitimacy of (some forms of) graffiti-writing also emerged which sought to provide opportunities for writers to develop their craft. Further, private economic interests began to sniff opportunities associated with graffiti – from advertising agencies mobilizing graffiti in efforts to reach the 'street' or 'youth' demographic, to publishing houses and aerosol paint manufacturers and suppliers, among others. As these interventions reshape the conditions for this form of public address, I have shown how graffiti continues to mutate.

As a form of public address, graffiti is quintessentially urban – graffiti writers have established quite particular ways of mapping and valuing the urban landscape in producing and circulating their texts to others. Indeed, as Halsey and Young (2002: 171) argue, 'reading a city's graffiti provides an alternative or anterior urban geography'. But this is not to say that the battles over the writing of graffiti have been contained to 'the streets' and other surfaces where it is written. As we have seen, the circulation of graffiti texts does not only occur by graffiti-writing upon a variety of mobile and static urban surfaces – graffiti texts also circulate on the pages of magazines, newspapers and books, and on television, cinema and computer screens. As such, struggles over graffiti-writing have ranged across various terrains. Certainly, attempts to eradicate (particular forms of) graffiti have mobilized barbed wire, surveillance and 'graffiti-proof' surfaces in order to prevent the unauthorized application of paint and ink to urban surfaces. But the war on graffiti has also involved a range of other strategies, such as attempts to regulate the sale of aerosol paints and computer games, and attacks on graffiti-writing magazines and websites – notwithstanding the usefulness of those very magazines and websites for urban authorities and social scientists who sought to gather knowledge of graffiti-writing cultures. Here, developments in Sydney reflect developments elsewhere. Authorities in other jurisdictions have similarly pursued graffiti writers across these various terrains – in the United Kingdom, public transport company Arriva has explicitly requested the banning of publications and computer games which feature graffiti-writing (London Assembly 2002), and some public libraries have blocked access to some graffiti-writing websites, in a move the authors of one of those websites described as 'virtual buffing'.[13] Struggles over the conditions in which graffiti is written, then, are not just struggles over the actual writing of graffiti-style *texts* on urban surfaces – they are also struggles over the circulation of graffiti as a *form* of public address.

In the context of this expansive war on graffiti, it is not surprising that the limits of the counterpublic horizon have been a source of considerable tension for graffiti writers. Graffiti-writing in Sydney relies upon the existence of a counterpublic sphere which facilitates both the circulation of graffiti texts and the development of graffiti-writing styles and skills. Regardless of whether the texts which circulate through this public sphere are intended only for other graffiti writers, as public texts they are potentially available to anyone – including police, journalists, youth workers, advertising agencies and social scientists (!). The existence of a graffiti-writing counterpublic sphere therefore makes graffiti both possible *and* governable. When knowledge of graffiti-writing circulates beyond graffiti writers, how will it be deployed, by whom, and with what effects for the ongoing practice of graffiti-writing? Efforts to 'keep it real', advocacy for legal 'aerosol art', and the commodification of graffiti can be understood as different responses to this tension – where the practice of illegal graffiti is sustained by a *tactical avoidance* of governance strategies, legal graffiti practices are sustained by strategic efforts to *shift* governance strategies and to establish market opportunities. These approaches, I want to suggest, are distinguishable by their different orientations towards the *audience* of graffiti texts.

Like the practice of beat-cruising discussed in the previous chapter, graffiti as *tactical avoidance* involves a complex combination of visibility and invisibility. While graffiti texts are visible to a wide audience on urban surfaces and in media, graffiti writers often try to maintain a kind of invisibility. Where the cover of cubicle doors and foliage is used by cruising men, graffiti writers use the cover of night and invented tag identities to evade capture. In media displaying their work, writers are usually clad in balaclavas or masks if they appear at all.[14] As Sydney writer Mistery put it, the graffiti writer seeks a kind of 'anonymous fame' (see *Basic Equipment*, 1998) – an identity which is simultaneously in wide circulation but 'anonymous' in the sense that it is distinct from their 'real' identity. Using their tag identity, writers are simultaneously able to address their work to the particular audience of other graffiti writers *and* to minimize the consequences of their texts circulating beyond that audience to hostile others. Indeed, from behind the cover of their tag identity, some writers have addressed this wider audience only to inform them that if they don't like graffiti, they can 'go and get fucked'.

The public sphere constituted by graffiti writers who write illegally and only for each other does not fit easily within conventional understandings of the public/private distinction. For example, in her study of graffiti in London, Nancy MacDonald (2001: 158) suggests that graffiti writers who operate behind the cover of a tag identity and direct their work only to other graffiti writers effectively 'band and bond together as a private and elite society':

> Writers use the city as their canvas aware that outsiders know nothing or little of the markings they see. This public yet very private parade of their subculture appears to give them a sense of power. The subculture is flaunted in the face of the public, but it remains out of their reach (MacDonald 2001: 158).

Based on a similar understanding of graffiti practices, Richard Sennett (1990: 205) was critical of 'graffiti kids' on the grounds that they were 'indifferent . . . to the general public, playing to themselves, ignoring the presence of other people using or enclosed in their space'. However, I think it is a mistake to characterize graffiti-writing as a form of private address, and it is certainly mistaken to critique this practice on the grounds of its 'indifference' to some imagined 'general public', as Sennett does. Writers who write for other writers are not so much establishing a *private* audience as drawing a sharp distinction between the *different public audiences* of graffiti – the desired audience of other graffiti writers, and the wider audience of non-graffiti writers. As I have argued, some differentiation of audiences is constitutive of all non-liberal counterpublic spheres. For Michael Keith (2005: 152):

> these outlaw forms of communication are precious because they transform the mainstream through writing over it, yet at the same time exclude the mainstream from their – the graffiti writers' – discourse.

Of course, those graffiti writers who draw such sharp distinctions between their audiences do not politically challenge the illegality of graffiti – indeed, the positioning of graffiti as deviant criminality by some urban authorities and mainstream media is reinforced by some writers who define authentic or 'real' forms of graffiti-writing by their illegality. Further, in their effort to maintain such sharp distinctions between audiences, some writers have effectively sought to police the boundaries of the graffiti-writing public by distinguishing 'insiders' from 'outsiders'. This strategy is an inversion of the very policing strategies which seek to position graffiti writers as anti-social outsiders in relation to the normative city/public whose values (including property values) they fail to respect (see also Keith 2005: 152).

In light of these limitations, some writers and their supporters have sought to establish legal graffiti opportunities. Here, tensions concerning the horizon of the graffiti-writing counterpublic sphere have been addressed by attempting to *shift* governance strategies which outlaw graffiti, rather than tactically avoiding them. Those who have sought and taken up legal opportunities for graffiti-writing have tended to blur the distinctions between the audiences of graffiti described above. When advocates for the provision of 'legal walls' have sought to shift wider perceptions of (some forms of) graffiti from 'vandalism' to 'art' or 'expression', their efforts have involved the circulation of public claims on behalf of 'aerosol art' to an audience beyond other graffiti writers. Further, the graffiti texts actually written on 'legal walls' are often designed in consultation with property owners and other interested parties, so as to be 'legible' (in the sense of being readable and/or simply attractive) to a wider audience. Similarly, many commercial opportunities for graffiti writers require legible graffiti texts when they are intended for a market only partially familiar with graffiti styles. But this engagement with wider publics also has its associated costs. The very notion of 'legal graffiti' conforms to established forms of authority sustained by the private property relation – indeed, legal graffiti reinforces this form of authority over the 'local authorities' established through illegal graffiti. Challenges to the property values which are the foundation of the city/public on whose behalf the 'war on graffiti' is waged are thereby foreclosed. And of course, the blurring of the distinctions between graffiti's audiences has a range of associated risks – not the least of which is that it threatens to undermine the invisibility which many graffiti writers have carefully crafted in order to avoid becoming a casualty in the war on graffiti.

More recently, a bridge between these two forms of graffiti practice – between illegal graffiti addressed to other writers and legal graffiti addressed to a wider audience – has begun to emerge in Sydney, and other cities in Australia and beyond. Artists frustrated with the constrictions of gallery spaces have begun to experiment with new forms of illegal and legal graffiti which are 'more visible, less cryptic and communicate to a wider audience' (Manco 2004: 8). In the work of these artists, there is an explicit attempt to use iconographic graffiti to provoke surprise and play in the urban environment, as a commentary on the encroaching corporatization and routinization of city life. Certainly, the likes of Richard Sennett may be pleased to see the emergence of these new forms of graffiti, with their explicit attempt to engage with the 'general public'. Indeed, these new forms are sometimes referred to by their supporters as *post-graffiti* (Manco 2004: 7) – the term first coined by Sennett (1990: 211) to describe those graffiti artists in New York and Paris whose practice 'evolved' from what he called 'simple smears of self' to more exploratory and experimental artistic practices which engaged with the urban fabric. In further echoes of Sennett, supporters of these new forms have sometimes sought legitimation through favourable comparisons with more 'traditional' forms of graffiti. A recent sympathetic examination of the blossoming stencil graffiti scene in Melbourne is illustrative:

> The National Gallery recently added specimens of stencil art on canvas to its permanent print collections. [The Gallery's print curator] sees stencil art as 'the most significant and vocal new trend in modern art in Australia'.
>
> On the other hand, 'tagging' – which refers to graffiti that often involves little more than crude lettering of the practitioner's street name – is like the ugly older brother of stencil art, and it is universally despised. Stencil art is distinct from traditional street graffiti which had its roots in the US hip-hop and criminal underworld. Most stencil practitioners consider themselves artists, many are women, and most baulk at being associated with organized crime. Stylistically, whereas traditional graffiti is made up of highly coded lettering, stencil graffiti is based on easily decipherable imagery (*Australian Financial Review*, 21 April 2006: 21).

It remains to be seen whether advocates and practitioners of these emergent graffiti practices can sustain their forms of address without resorting to this tired distinction between 'art' and 'vandalism'. After all, this very distinction has characterized three decades of efforts to limit opportunities for graffiti-writing in Sydney on behalf of a fictitious 'general public'.

My point here is not that we should sidestep the need to make judgements about the quality and impact of different forms of graffiti, simply celebrating all graffiti for its 'transgressive' properties. Rather, in this chapter I have sought to show that the struggle to produce a graffiti-writing counterpublic sphere is also the struggle to construct and circulate alternative, non-liberal frameworks by which graffiti can be judged.

Chapter Six

'no fun. no hope. don,t belong.': Remaking 'Public Space' in Neo-liberal Perth

This is the first of two chapters in which I want to approach the urban dimensions of public-making from a slightly different angle. My focus is on two efforts to secure space on behalf of particular forms of public sociability through the exclusion of particular people who are conceptualized as a threat to those forms of sociability. Perth, the capital city of the state of Western Australia, is a fitting place to consider in this context. It has become the first city in Australia to formally decree a night-time curfew for young people. Launching the policy in June 2003, the State's Premier Geoff Gallop (2003a) said:

> This is about protecting children who, quite frankly, should not be wandering the streets at night . . . It is also about protecting the rights of people to go about their business in Northbridge without being harassed by gangs of juveniles. In many cases they are engaging in aggressive and offensive behaviour making them not only a nuisance to others but a risk to themselves. Many are under the influence of alcohol or other drugs and in obvious physical and moral danger.

Perth is by no means unique in establishing a curfew for young people (Collins and Kearns 2001). If mainstream media and political rhetoric in many cities is to be believed, then one of the spectres haunting 'our' streets is groups of 'anti-social' teenagers. These teenagers 'wander' and 'roam', they wear hooded tops and baseball caps, and they 'threaten' and 'intimidate' members of 'the community'. Even if they are not engaged in crime, they are up to no good. In cities across the United States, the United Kingdom, and Australia, extraordinary efforts are being directed to developing new ways of dealing with this 'anti-social behaviour'. And while there are important differences in both the rhetoric and practice of addressing 'anti-social behaviour' in these cities, there is also clear evidence that ideas are circulating between these cities and helping to establish some common responses. The widespread diffusion of the 'broken windows' theory of community safety – from the coining of the term in 1982 by criminologists Wilson and Kelling (1982; see also Kelling and Coles 1997), to its popularization in New York City under Mayor Giuliani and spread throughout the United States in what Smith (1998) describes as the 'revanchist 1990s', to its repetition in speeches by British Prime Minister Tony Blair (2001) and his government's 2003 White Paper on Anti-Social Behaviour (entitled *Respect and Responsibility: Taking a Stand*, see Home Office (UK) 2003) – is one particularly powerful example of such circulation.

In such rhetoric, anti-social behaviour is identified as a key obstacle to achieving 'quality of life' and 'liveability' of cities. As former UK Home Secretary David Blunkett put it in his Ministerial Foreword to the Anti-Social Behaviour White Paper:

> The anti-social behaviour of a few, damages the lives of many. We should never underestimate its impact. We have seen the way communities spiral downwards once windows get broken and are not fixed, graffiti spreads and stays there, cars are left abandoned, streets get grimier and dirtier, youths hang around street corners intimidating the elderly. The result: crime increases, fear goes up and people feel trapped (Home Office (UK) 2003: 3).

Streets and other *public spaces* figure highly in quality of life/liveability agendas. In more combative rhetoric, such as that mobilized by New York City's Mayor Giuliani and Police Commissioner Bratton in their (in)famous 1994 *Police Strategy No. 5: Reclaiming the public spaces of New York*, public space has to be taken back on behalf of decent citizens:

> New Yorkers have for years felt that the quality of life in their city has been in decline, that their city is moving away from, rather than toward the reality of a decent society. The overall growth in violent crime during the past several decades has enlarged this perception. But so has an increase in the signs of disorder in the public spaces of the city . . . Mayor Rudolph W. Giuliani has called these types of behaviour 'visible signs of a city out of control, a city that cannot protect its space or its children' (quoted in Chesluck 2004: 256).

The 'improvements' to 'public space' made under the rubric of quality of life and liveability agendas have of course attracted a good deal of critical attention. Critics have been particularly concerned with the exclusionary consequences of these policy agendas for groups who are declared to be 'anti-social' rather than part of the 'community' – such as the homeless (for example Davis 1990; MacLeod 2002; Mitchell 1997) and young people (for example Valentine 1996; White 1996). These critical analyses of 'public space' in contemporary cities have increasingly come to be framed with reference to wider discussions concerning the emergence of 'neo-liberal' approaches to urban governance (for overviews of these discussions, see Peck and Tickell 2002; Brenner and Theodore 2005). Like other 'public services', it is feared that 'public space' has fallen victim to the prevailing neo-liberal ideologies of 'privatization' and 'public–private partnership'. While the historical liberal concept of urban public space may only have valued 'inclusion' as a means to civilize the masses by bringing them into contact with bourgeois society, the neo-liberal approach to public space is said to have jettisoned such values in favour of outright hostility towards the less well-off:

> The old liberal attempts at social control, which at least tried to balance repression with reform, have been superseded by open social warfare that pits the interests of the middle class against the welfare of the poor (Davis 1992: 155).

Critics of neo-liberalism worry that the governance of urban populations through the normative gaze of their peers has given way to a territorial separating-out of the 'community' from the 'anti-social', establishing disconnected 'circuits of inclusion' and 'circuits of exclusion' in contemporary cities (Rose 2000). Those in the circuits of inclusion move through an archipelago of safe spaces which are secured for the 'community' against the 'anti-social'. The anti-social, on the other hand, are banished to peripheral spaces, or have their movements restricted through the imposition of 'good behaviour contracts' and incarceration.

In examining the dynamics of Perth's city centre transformation, it is tempting to ask whether or not the patterns of exclusion in the paradigmatic places of neo-liberal urban governance (such as Davis's Fortress Los Angeles and Smith's revanchist New York City) have been replicated there. Are the new permutations of capital accumulation and social control pioneered in those cities spreading to other cities around the world, even to Perth? According to Rob White (1996: 47), Australians have 'much to learn' from general analyses of neo-liberal globalization, and from specific analyses of cities like 'Fortress LA':

> The main message of these kinds of study is that deep social inequality and social division is increasingly being accompanied by the militarization of the landscape (i.e. through extensive 'fear' campaigns and the fixation with 'security' at all costs); by the active creation of 'social control districts' based upon containment strategies (e.g. those designed to quarantine social problems and 'problem' populations); and by the parallel processes of ghettoization and fortification of neighbourhoods.

In Australia, and elsewhere, invoking the experience of 'Fortress LA' or some other paradigmatic place as a cautionary tale is one way for writers and planners to make a dramatic case for change, lest 'we here' end up like 'them over there'. Mark Peel (1995: 40) notes that 'what confirms the point are words and pictures of urban inequality, LA-style: gates, walled enclaves, video surveillance'. All of these features are present in contemporary Perth, and more besides.

However, others have urged caution in taking this approach. The 'neo-liberalization of space' is neither uniform nor unfettered in cities across the world, even where neo-liberal ideology and discourse may be observable in political rhetoric and interventions. In the Australian context, O'Neill and Argent (2005: 5) suggest that neo-liberalism is always *contingent*, its form and its outcomes not pre-determined with reference to some model derived from experiences elsewhere. For them, a ' "contingent" neoliberalism means that, while we can observe neoliberalist tendencies, its material form can never be predicted or guaranteed'. To fully appreciate this contingency, Larner (2003: 510) suggests that:

> we need a more careful tracing of the intellectual, policy, and practitioner networks that underpin the global expansion of neoliberal ideas, and their subsequent manifestation in government policies and programmes.

So, to critically assess the remaking of 'public space' in Perth, it is not enough to ask whether elements of 'Fortress LA' or some other paradigmatic place have been replicated in Perth. The multidimensional approach to publics and the city that I have developed in this book offers one useful means to respond to Larner's call for close analysis. It does so by emphasizing that we cannot take topographical models of 'public space' for granted, and so we must look beyond what has happened *in* Perth's 'public spaces' to investigate how 'public space' has been made an object of public action, and to critically interrogate the model of 'the city' which underpins such action. In this chapter, I begin by drawing attention to the ways in which the revitalization of 'public space' became an object around which the activities of a loose coalition of actors in Perth's urban restructuring were organized. As we shall see, the activities of this coalition range across a wide variety of spaces, and were the product of a shared vision of the interests and needs of the Perth's city centre as a 'Capital City'. The design and implementation of governance measures to realize this vision of Perth's city centre produced a series of contradictions and challenges. The rest of the chapter explores the contests over how these challenges were to be met. In particular, I focus my attention on the emergence of a widely shared concern about (some) young people's occupation of street spaces in the city centre. Efforts to address this concern have taken a variety of forms, not only attempting to deny young people access to the newly restructured 'public spaces' of the city centre, but also positioning young people outside of 'the city' in a symbolic sense. The 'city centre' has come to be defined explicitly as an 'adult space', and the 'proper place' of young people has been fixed in 'the suburbs' and 'the family'. The chapter then considers a variety of efforts to contest this vision of the city centre. In considering these efforts, I assess the grounds on which alternative governance aims and mechanisms were proposed and implemented by critics of the governing coalition. Finally, I offer some concluding reflections on the implications of Perth's restructuring for our understanding of neo-liberal forms of urban governance.

'City Visions': Promoting the Revitalization of 'Public Space' in Perth

During the late 1980s and early 1990s, a coalition of actors came together to promote a new identity for Perth's Central Area. The construction of a common conception of 'public space' was particularly important in establishing a common cause among the diverse actors in this coalition. The revitalization of public space was generally agreed to be of central importance in (re-)establishing the Central Area as a distinct, discrete locus of a range of urban functions and practices. Three agendas for urban restructuring being pursued by different interests in government and the private sector all came to attach significance to public space revitalization – the desire to establish Perth as a genuine *Capital City* for the State of Western Australia, the desire to increase *inner city living*, and the desire to bolster *law and order* in the city centre. I now want to consider each of these agendas in turn, showing the ways in which ideas about public space were generated and circulated in the initial phase of Perth's contemporary restructuring.

The 'Capital City' agenda

In 1988, a small group of local planners, academics and architects formed a lobby group to focus attention on the planning of Perth's centre. CityVision, as they called themselves, never involved more than 25 people. The manifesto they published in 1988, *New Directions for Central Perth*, was not the first document to identify a need for reform to the planning of the Central Area – others had voiced similar concerns in the 1960s and 1970s (see for example New Heart for Perth Society 1969; Stephenson 1975). Nonetheless, CityVision's agenda received increasing amounts of attention in the wider public sphere in the years following its publication (see press clippings in Australian Institute of Urban Studies WA Branch 1992). CityVision argued that the inner city had become a lifeless, concrete jungle. They urged governments to adopt policies which they considered would revitalize the Central Area, and suggested the adoption of the following nine key planning and design principles to achieve this:

- restore diversity and vitality to the city;
- bring residents back to the city;
- make the city an enjoyable place, day and night;
- make the city a pleasant and stimulating place to be in;
- make the city accessible to all people;
- seek excellence in urban design and preserve our heritage;
- bring together the city and the river;
- involve people in the future of their city;
- establish a planning and development system designed to realize the full potential of Perth as a true capital city (CityVision 1992 (1988)).

CityVision's manifesto struck a chord with the WA Chamber of Commerce and Industry (WACCI), which had itself begun expressing concerns over the future of Perth's Central Area. In particular, business interests represented by WACCI (1992: 27) felt that there was:

> the lack of balance between the CBD as the main provider of revenue and its lack of representation in a council with a strong pecuniary interest in maintaining the status quo.

A new planning approach was needed, they argued, to ensure that the Central Area of Perth was rightly treated as a distinct locality – a 'Capital City'. The design initiatives proposed by CityVision could contribute by giving this locality a distinct urban identity.

The 1990 *Capital City Forum*, jointly organized by WACCI, the WA branch of the Australian Institute of Urban Studies and CityVision, gave this nascent coalition a chance to put their proposals to government planners, politicians, and the media. The Forum resolved in favour of the creation of a Capital City Planning Authority, which would be appointed (not elected) by the State Government and involve equal numbers of government, business and community representatives. They also proposed that the Perth City Council (PCC) boundaries be redrawn to exclude the suburbs and focus solely on the revenue-raising CBD. Not surprisingly, the Lord Mayor of Perth at the time was not impressed (*West Australian*, 21 September 1990: 8). The PCC commissioned a consultant to report on the feasibility of establishing a Capital City Authority, and the report predictably and firmly rejected the merits of the plan (*West Australian*, 21 February 1991: 12).

Notwithstanding these emergent tensions, the Capital City agenda began to gain momentum in the early 1990s. The State Labor Government began to conceive of its own interest in giving the inner city a new identity, in order to attract investment. Acting Premier Ian Taylor told the *West Australian* (15 September 1990: 9) after the *Capital City Forum* that 'future investment capital was going to be very choosy about where it invested' and 'reform of the city and its relationship was a vital ingredient' in attracting it. The redesign or revitalization of public space figured centrally in the Government's thinking, and this was reflected in its 1990 *Metroplan* (Department of Planning and Urban Development (WA) 1990). This vision document argued that Perth's Central Area should be given a new identity through projects which gave the centre 'an identifiable built form' (Department of Planning and Urban Development (WA) 1990: 46). More cultural and leisure facilities, more residential and tourist accommodation, and improvements to public space were all conceived of as means to 'strengthen Central Perth's role as the State's premier centre' by adding to 'the character and vitality of the Central Area' (Department of Planning and Urban Development (WA) 1990: 44). Metroplan also raised questions of urban governance in Perth, drawing on the Capital City concept in order to suggest changes to planning structures for the Central Area (Department of Planning and Urban Development (WA) 1990: 45).

The Perth Central Area Policies Review Study Team was formed to further explore issues raised in *Metroplan*, and their report was submitted to the State Government and the PCC in 1993. Their document identified five 'City Challenges' for Perth's Central Area, expanding on the issues raised in *Metroplan*. Public space design and changes to governance figured significantly in all five:

- ensuring a thriving, prosperous and economically competitive city;
- enhancing the city centre's character and charm;
- creating a vibrant city centre for enjoyment, living and learning;
- making the city accessible;
- managing for a dynamic city centre (Perth Central Area Policies Review Study Team 1993).

To meet these challenges, the city centre had to be 'redefined', so that it no longer relied on a narrow base of retail and administrative activities. Such a narrow base was said to be unsustainable given the new forces shaping the city, such as: instant worldwide communications; global economic rationalization; and information as the basic industrial commodity (Perth Central Area Policies Review Study Team 1993: 17). Most importantly, the revitalization of the city centre's public spaces was necessary both to lift the profile of Perth in the Asia-Pacific region, and to ensure that the

Central Area was 'the juxtaposition of suburban sameness' (Perth Central Area Policies Review Study Team 1993: 18).

This discourse of city 'challenges' was a direct reflection of the 1992 *City Challenges* conference, which had been co-sponsored by the State Government and the PCC in a further effort to build political momentum and raise the public profile of their concerns about Perth's Central Area. The conference was strongly supported by personalities associated with CityVision, who were responsible for inviting Copenhagen-based urban designer Jan Gehl to present ideas for the reinvigoration of Perth's public spaces. He promised conference delegates a relatively simple set of solutions to Perth's problems:

> In 20 years, the number of pedestrians using Copenhagen has increased 25 per cent even though the population and tourist through-flow have remained static. Also, the activity level by people in the city streets has increased three fold. Perth can achieve similar results by adopting a simple formula – one that has many similarities with procedures needed to plan a successful party within the home (Gehl 1992: 29).

The State Government and the PCC were so impressed that they invited Gehl to undertake an extended study of Perth's public spaces. This study began in 1993, supported by a steering group involving planners from the PCC, the State Government and members of CityVision.

Two issues particularly troubled Gehl. First, he attacked the mono-functionality of the network of precincts into which Perth's Central Area was divided. Like Jane Jacobs (1961), he emphasized the importance of multifunctional public spaces for vitality and safety:

> This unusually rigid subdivision of functions is most unfortunate for the creation of a lively, diverse and safe day and night city. Researchers into urban activities as well as crime prevention and safety all point to the necessity of a good mixture of functions (Gehl 1994: 14).

Second, Gehl argued that the potential for 'higher quality walking – or promenading – from one end of the city to the other is not provided' (Gehl 1994: 4). Rather, streets tended to be dominated by the car:

> In Perth one can see advertisements (for car parks) saying: 'In Perth your car is as welcome as you are'. That the cars are most welcome is evident but certainly you yourself are not, from the moment you leave your car, outside a few select areas (Gehl 1994: 24).

Gehl's recommendations for improvement centred on creating public spaces in which a pedestrians might undertake not only necessary activities, but also optional activities and social activities. The recommendations bore a striking resemblance to the original CityVision manifesto, focusing on the need to:

- capitalize on Perth's unique qualities (particularly its natural setting and climate);
- expand the heart of the city;
- reduce car traffic in the city;
- bring the city and river together;
- create a better city to walk in;
- create a better city to stay in (particularly by creating a new network of squares and sidewalk cafés);
- create a lively and safe city (particularly by increasing the number of residents).

This last recommendation concerning the residential population of Perth's Central Area also draws our attention to another agenda for restructuring of Perth's Central Area which was beginning to gather pace alongside the 'Capital City' agenda – inner city living.

The 'inner city living' agenda

The redefinition of public space in the Central Area was pushed along by the State Government's desire to increase the residential population of the inner city. Both the *Central Area Policies Review* and *Metroplan* had argued that there were a number of environmental and economic reasons for increasing urban densities in Perth. The Central Area, it was hoped, could play a significant role in contributing to the provision of new dwellings in established urban areas to stop further suburban 'sprawl'. This agenda happily coincided with the belief of CityVision, Gehl and others that more inner city residents would add to the vitality and safety of the Central Area.

An Inner City Living Taskforce (hereafter 'the Taskforce') was established, and their 1992 report was framed by an acceptance of the prevailing logic that additions to the inner city's housing stock would reverse the decline of the Central Area's population, thereby contributing to a city which would be 'more efficient, economical and environmentally sustainable' (Perth Inner City Living Taskforce 1992: 17). The Taskforce had quite a job ahead of them. Between 1954 and 1985, the number of dwelling units in the Central Area had dropped from 7,000 to 1,600, and the population had declined by over 60 per cent, with only 5,600 residents remaining (Perth Inner City Living Taskforce 1992: 18). According to the Taskforce, the 'urban village' would be the Central Area's saviour. In order to encourage the development of inner city urban villages, the State Government and PCC had three key objectives:

- to demonstrate the benefits of inner city living;
- to promote the inner city lifestyle; and
- to facilitate the provision of inner city housing by private developers (Perth Inner City Living Taskforce 1992: 27).

Improvements to public space seemed to contribute to each of these objectives. Government planners believed potential residents and developers would consider the Central Area a desirable residential location if it had a unique identity. The Taskforce enthused that:

> The magic of the city is due to the intensity and density that comes from the grouping of many activities and functions in the one place . . . The Urban Village has the potential to provide the mix of activity, the excitement, the diverse, yet cohesive, physical environment which has that very important element: sense of place (Perth Inner City Living Taskforce 1992: 24, 26).

The urban village concept had already been applied to a major urban renewal project for a 120-hectare waterfront site at the east end of Perth's Central Area, which had been unveiled by the State Government in 1990. According to the Government, declining land values in the area associated with de-industrialization made it ripe for redevelopment into a new urban residential precinct (Government of Western Australia 1990). The East Perth Redevelopment Authority (EPRA) was established as a statutory authority by legislation in 1992, part-funded by the Commonwealth's Better Cities program, to assemble land, provide infrastructure, and plan and promote the overall development of East Perth. The EPRA's role was to be limited to 'traditional roles of land assembly, provision of infrastructure, planning, promotion and overall development management' (Government of Western Australia 1990: xii). Housing would be provided by private developers, to be developed in line with objectives and standards established by the EPRA.

The law and order agenda

Concerns with youth crime and 'anti-social behaviour' were a central feature of discussions about the remaking of public space in the Central Area. In August 1991, a 'Rally for Justice' was held outside Perth's Parliament House. Promoted by talkback radio 'shock jocks', the rally attracted an estimated 20,000 people, and aimed to pressure the State

Government to introduce tougher sentencing laws for juveniles, particularly targeting repeat offenders and car thieves (Mickler and McHoul 1998: 129). This rally framed an intense media debate over youth crime in the following months of 1991, despite a continuing decline in the actual numbers of car thefts recorded in Perth. This media debate about 'youth' crime had a particular inflection. As Howard Sercombe (1995) found in his study of references to young people in the main Perth newspaper, the *West Australian*, between April 1990 and March 1992, 'the face of the criminal is Aboriginal'. Just over 10 per cent of all articles about young people across the study period made direct references to Aboriginality, outnumbering the next largest reference to ethnicity by ten to one. Of these references to Aboriginality, 85 per cent of articles were principally about crime. In a similar study of the WA print media, Mickler and McHoul (1998: 130) found that in the months of May, September and November 1991, and January 1992, youth-crime reports identifying Aboriginality made up between 60 and 95 per cent of all youth crime reports. Such statistics bore no direct correlation to the actual number or proportion of crimes committed by Aboriginal young people.[1]

The Labor State Government responded in 1992, introducing new sentencing laws for juveniles through the *Crime (Serious and Repeat Offenders) Sentencing Act 1992* (Mickler and McHoul 1998: 122). However, this was not enough for the Opposition, whose 1993 State election campaign focused on Labor's failure to address juvenile crime in Perth, particularly offences such as car theft and graffiti. Their policy statement *Project Perth – A Living and Working Capital* (to which we will return shortly) argued that:

> Perth is in the grip of a crime crisis which our police force is powerless to stop . . . Labor's soft approach to the state's growing crime wave has only made matters worse (Liberal Party (WA) and National Party of Australia (WA) 1993: 10).

Opposition leader Richard Court promised to 'get tough' by giving police the extra powers and resources they needed to fight this 'crime crisis'.

The PCC had already responded to media-fuelled perceptions of a crime wave with the installation of the first and largest closed-circuit surveillance camera network of any Australian city. In 1991, it placed 48 cameras in the central shopping precinct, covering Hay and Murray Street malls, as well as Forrest Place and the central railway station (City of Perth nd). The expense of installing and operating the closed-circuit television (CCTV) network in the Citiplace retail precinct (AUS$750,000 for installation, AUS$250,000 annually for operation) was justified on the grounds that it would protect the state and private sector investments in public space. The

Figure 6.1   Street life under the eye of a CCTV camera, Perth Central Area (photo: Kurt Iveson)

PCC (City of Perth nd: 1) had found that 'the best efforts of the Council to rejuvenate the area were often offset by emotional stories in the media which generated an undesirable image of the City', and this had meant that 'a section of the residential population dislikes visiting the central city for reasons of insecurity'. With the installation of the CCTV network, efforts to envision the public spaces of the Central Area were bolstered by the ability to actually see what was happening in these spaces, and by the PCC itself being seen to be doing something (see Figure 6.1).

The emerging vision of Perth as a City/Subject

In producing their representations of public space, the various actors involved in the discussions staged over the 1980s and early 1990s both drew upon and contributed to a wider conception of the 'City of Perth' as a particular kind of city. The remaking of the 'City' was not simply a matter of refashioning the city's built form, it was also a political project which sought to establish the City as a *subject*, with its own interests, identity and needs (on the city as subject, see Keith 2001). Those engaged in the product of producing Perth as a City/Subject conceptualized their task as addressing a number of 'challenges' to the City's status and progress. Means had to be found to give the Central Area a 'capital city-ness' that would establish Perth as an exciting, urban destination for visitors from overseas, inter-state and the suburbs alike. The respective roles of the state and 'the market' were clearly articulated in the agenda for the Central Area – various state agencies were concerned to establish the governmental structures, development controls and infrastructural improvements (to areas such as transport and 'public space') in order to encourage 'private' investment in housing, retail and other employment. The state could make a contribution not only to vitality and excitement, but also to safety and convenience.

This emerging vision of the city of Perth survived a change of government in 1993, with the election of Richard Court's conservative Coalition Government. Certainly, there were some modifications to various policies – but none of these modifications radically challenged the vision of 'the City' which had taken shape under Labor. The new State Government's agenda for the Central Area was unveiled in the 1995 report *Perth – A City for People* (Government of Western Australia and City of Perth 1995). Specific design projects were proposed to enhance the 'identity' and 'vitality' of the various precincts – all accompanied by ink and watercolour impressions full of happy people using the newly refurbished public spaces. The following introductory passage from *Perth – A City for People* captures the continuing importance of public space as a key ingredient of the new State Government's vision for the Central Area. While Perth had already established a good name:

> for those who look past this well deserved reputation and examine the city at ground level, it presents a picture of many unused opportunities. The challenge before us is to turn those opportunities into features which lift the city to new levels of excitement, colour, convenience and comfort for millions of visitors and for those who traverse its streets each day for work, shopping and entertainment (Government of Western Australia and City of Perth 1995: 1).

As we shall see, Court's Government took action on a number of fronts to advance the 'Capital City' agenda during its years in office.

City Challenges Pt 1

I now want to trace a series of unfolding efforts to materialize the city visions discussed above, through a range of interventions designed to bring this new City into being by attaching a new identity to the Central Area. I focus particular on interventions in the realms of urban governance, urban design, social planning and spatial regulation. As we shall see, as the various agendas for urban restructuring attempted to address existing urban 'challenges', they also provoked new challenges and difficulties for urban authorities which threatened to destabilize the coalition of interests that had formed in the name of Perth's Central Area.

Governing the Central Area

Perhaps the Court Government's most decisive action regarding the Central Area was to break up the Perth City Council. Consultants were engaged almost immediately after the change of government to provide a report to Premier Court about the governance of the Central Area. According to the report, the existing council boundaries meant that 'the heart of the City – the Central Area – has not received the attention deserving of its role in the Perth Metropolitan Region and the State of Western Australia' (Carr and Fardon 1993: 2). As a consequence:

> The 'Heart' of the State is in trouble – vacant offices and shops, increasing blight and crime, exodus of executives and workers to the Eastern States and Regional Centres and declining property values are reasons for growing concern (Carr and Fardon 1993: 3).

The report recommended that it was time for the PCC to be restructured. Key requirements for the new system included:

- Central Area interests to have a greater say in decision-making for the Central Area;
- a better, stronger and properly resourced co-ordination mechanism between the State Government and the Perth City Council;
- better mechanisms and processes to involve key interest groups in the planning and development of the Central Area;
- the marketing and promotion of the attractions of the Central Area (Carr and Fardon 1993: 5).

The consultants proposed that the suburban residential areas of the PCC be separated into three separate municipalities, and that the Central Area be governed by a newly reconstituted Perth City Council, in consultation with the State through the newly established Premier's Capital City Committee. These recommendations bore a striking resemblance to proposals developed by the organizers of the Capital City Forum in 1990. The existing PCC was dissolved, and Commissioners were appointed, with elections to be held for the new PCC in 1995.

Designing and marketing the Central Area

Informed by the general principles set out in *Perth – A City for People*, the State Government and the newly restructured PCC set about conducting a series of studies into various aspects of the Central Area, from lighting and furniture to the needs of specific precincts and population groups (see for example KPMG 1995; Government of Western Australia 1995; City of Perth 1995a, 1995b, 1995c; Planning Group Pty Ltd and Godfrey Spowers Puddy and Lee 1995; Government of Western Australia and City of Perth 1996). Not everything proposed within these various reports actually came to pass. But a variety of projects involving public space enhancement – some jointly funded by the State Government and the PCC through *Perth – A City for People* – did take place, including: the flagship refurbishment of King Street; widening of footpaths, improvements to Russel Square, and the conversion of old council depots into medium density housing in Northbridge; new seating and shading for Hay and Murray Street Malls; substantial public investment in new urban design at East Perth; substantial renovation of the central railway station and nearby Cultural precinct; the introduction of the free buses in the Central Area. After this initial suite of improvements, the various discrete strategies were drawn together in the 1997 *Public Space Enhancement Strategy* (Government of Western Australia and City of Perth 1997). These efforts began to pay dividends, with substantial investment in housing and commercial buildings also approved, such that the *West Australian* (10 March 1997: 15) was happy to report in 1997 that 'for the first time in many years, construction cranes can be seen again on the skyline'.

Both the State Government and the PCC were keen to promote and protect their substantial investment in the redefinition of Perth's Central Area. The audience for improvements to the Central Area had always been international and local. In November 1996 'supermodel' Elle Macpherson was engaged to appear in a series of commercials promoting tourism in WA internationally in a controversial deal that would cost WA Tourism $7 million (*West Australian*, 2 July 1997: 1). At a less extravagant scale, the PCC launched its *See You in the City* campaign, involving billboard, print media and television commercials to Perth residents (see Figure 6.2). This campaign was primarily designed to attract shoppers and potential residents from the suburbs. Of course, both the slogans of *Perth – A City for People* and *See You in the City* begged the question: who would the Central

Figure 6.2 'See you in the city' promotional signage, Perth Central Area (photo: Kurt Iveson)

Area be for? As the design improvements and marketing began to roll out, tensions over this issue began to surface.

The 'social mix' of the Central Area

Visions of inner city housing and urban villages had typically made reference to the valuable 'social mix' of inner city communities, which could contribute to their vitality and growth. But the question of how this mix was to be achieved proved a difficult one for city planning agencies and developers alike. The Labor Government's Inner City Living Taskforce (1992: 5) had sought some commitment to the provision of low-cost housing in the Central Area in order to achieve the desirable 'mixed uses and social mix'. However, both the State and City authorities failed to act on (or commit to) this aspiration in any effective manner. A survey of intending inner city residents conducted in 1994 did find a high level of demand for low-cost housing. However, the report on the survey simply concluded that 'it will be difficult to satisfy demand in the lower end of the market, due to the nature of land prices in the city' (Department of Planning and Urban Development (WA) 1994: 3). The Government was more concerned to relax planning codes for private developers, implementing new standards based on 'a more flexible and innovative approach to the development of housing in inner city locations' (Planning Group Pty Ltd and Godfrey Spowers Puddy and Lee 1995).

The PCC was similarly uninterested in actively shaping the social mix through the provision or protection of low-cost housing in the Central Area. When a developer proposed to knock down existing blocks of flats on Burt Way near East Perth to make way for a cluster of luxury buildings, including a fifteen-storey apartment tower, residents mobilized to oppose the development. The Lord Mayor of Perth Dr Peter Nattrass said he had nothing against the existing residents, but nonetheless found himself in a difficult position:

> You have got to weigh up the rights of a person who has taken six years to buy up all this property. We are not allowed to make social decisions (quoted in *West Australian*, 4 September 1997: 7).

Notwithstanding Nattrass' reticence to make 'social decisions', the redevelopment proposal was eventually withdrawn. Residents, however, felt that their victory represented only a small hiccup in the longer-term decline in the stock of affordable rental housing in the Central Area (Perth Inner City Housing Association 1997a).

The lack of any effective will to intervene in the housing market to supply low-cost housing was compounded by the State and PCC's urban design improvements to public space in selected parts of the Central Area, which tended to provide a significant boost to housing prices. The improvements to King St provide one small example. The PCC and the Coalition Government spared no expense in giving King Street a new identity, significantly investing in improvements to the quality of public space and constructing a new AUS$4.3 million arts centre. The redevelopment of King Street significantly boosted land and property values, and a new development of 'luxury serviced apartments' taking in three heritage listed buildings capitalized on the transformation. Potential buyers were told:

> King Street was made famous by the rag trade in years gone by and today little has changed with the cobblestone street a haven for chic retail outlets like Louis Vuitton, Cartier, King St Couture and many other exclusive and fine establishments. Trendy restaurants and coffee houses make this a most desirable area in which to live (*West Australian*, 1 March 1997: Real Estate Liftout, 57).

The successful redevelopment of King Street had, if anything, helped to reduce the social mix. Further, according to some PCC councillors, heritage regulations had been side-stepped in the approval of some developments (*West Australian*, 27 February 1997: 3).

The story was similar in East Perth, although on a much larger scale. The EPRA spared little expense in the urban design of public spaces in East Perth in an effort to attract private developers to invest in the area. The developers came, and not surprisingly, apartments and houses in East Perth began to attract very good prices. While in 1996/97 prices all over Perth had begun to increase, the Real Estate Institute of WA reported that East Perth prices had 'outstripped the best of them with a 17.5 per cent increase in prices over the past year' (*West Australian*, 30 July 1997: 11). The EPRA had made a gesture towards the provision of some low-cost accommodation in East Perth, determining that 50 units should be set aside in the area for public or community housing purposes. An arrangement for the provision of the first of these units – a total of 19 in the Haig Park development – was finally negotiated by the Perth Inner City Housing Association in 1997 (Perth Inner City Housing Association 1997b). To date, a further 37 units of public housing have been provided in Perth, making a total affordable housing component of 56 units. While some of the funds spent by the EPRA were provided by the Commonwealth Better Cities programme, the Commonwealth simply failed to establish any accountability mechanisms with regards to housing affordability (Crawford 2003).

Tensions over the social mix of inner city housing were not restricted to questions of affordability. Peter Murphy and Sophie Watson (1994: 587) have noted that as Australian cities increasingly look outwards for capital investment, the closest source of this investment has been the Asia-Pacific region. The influx of tourism and other forms of investment from this region has sparked tensions and racist concerns over the nature and amount of 'Asian influence' in Australian cities. Such tensions were apparent in Perth. While by 1997 sales of Central Area housing had begun to grow, the dominance of overseas (predominantly Asian) investors in the inner city property market attracted increasing attention. Developers, the EPRA and the PCC made a renewed pitch to local residents. In a jointly funded six-page advertising feature on 'Inner City Living' in the *West Australian*, these organizations made clear the 'preferred' occupants for the 'resurgence of residential accommodation within Perth City':

> Many of these developments have attracted overseas interest but it is the owner-occupiers who most developers want to see filling the space (*West Australian*, 9 July 1997: Real Estate Liftout, 51).

This desire for owner-occupiers tended to push the prices of residential apartments up even further, because only the most expensive housing seemed to appeal to the 'empty-nester' from the suburbs, while smaller and cheaper apartments were purchased by (often international) investors (*West Australian*, 9 July 1997: 57). The developers of the Mounts Bay Village assured readers that 'two thirds of all the residential and serviced apartments . . . have been sold off the plan to West Australians' (*West Australian*, 9 July 1997: Real Estate Liftout, 52). Not all developers were quite so concerned about the ethnicity of potential buyers. Some pitched more directly to the Asian market – for example, investors from Asia buying properties for their children studying at Perth universities were particularly welcome at Burswood Gardens, which was in 'close proximity to universities' and where 'half the development has been sold to overseas interests' (*West Australian*, 9 July 1997: Real Estate Liftout, 56). The Paramount made a play for both markets, hoping to catch the locals with the promise of cappuccino from the downstairs café, while also enticing the overseas investor: 'it's an ideal residence for overseas residents attending tertiary schooling and English or business studies in Perth' (*West Australian*, 9 July 1997: Real Estate Liftout, 55).

Whoever the new residents turned out to be, they were to be spared from other outbreaks of diversity in the Central Area's emergent residential landscape. Lord Mayor Natrass was not prepared to relinquish all of his capacity to make 'social decisions', particularly on behalf of the families he hoped would move back to the inner city. In December 1996, the PCC put a blanket ban on amusement arcade developments in the city's central shopping precinct, and decided it would approve new arcades in other precincts only if operators agreed to follow a code of conduct to 'stop anti-social behaviour' (*West Australian*, 12 March 1997: 73). Nattrass intervened to prevent a methadone clinic from being relocated to an address in East Perth near the Royal Perth Hospital, on the grounds that 'East Perth is undergoing rapid transformation, becoming an attractive residential area' (*West Australian*, 5 July 1997: 42). The PCC also rejected an application by Northbridge adult shop Club X to run adult video booths, with Nattrass quoted as saying that the PCC wanted to 'encourage families to come to Northbridge' (*West Australian*, 10 July 1997: 36). The introduction of CCTV surveillance by the PCC had been designed to make the Central Area 'a pleasant place in which families can gather to enjoy shopping and participate in the many and varied public events that occur in this area' (City of Perth nd: 7).

Securing the Central Area

Concerns about law and order in the Central Area, already in full swing in the beginning of the 1990s, gathered momentum with the growth of inner city residential development, and provoked a wide range of interventions – from design measures to new policing strategies. Planners had initially hoped that security would take care of itself, through a combination of improvements to design (such as lighting) and the greater number of 'eyes on the street' which would result from a larger residential population (Perth Inner City Living Taskforce 1992; see also Gehl 1994). After all, this is what seemed to happen in those European cities idealized in the urban village and city living literature:

> It is the European City which first gave us the street and the city square – based upon the simple principle of the front of the house looking out on to the outside world and public activity: the back of the house facing inwards giving the opportunity for intimacy and privacy. These streets not only produced beautiful areas, they created spaces which stimulated social responsibility (Perth Inner City Living Taskforce 1992: 60).

But this ideal ran up against the need to provide more security (and more parking spaces) to entice reluctant suburbanites to move to the city.

State Government surveys of Perth residents and those who had registered as 'intending residents' of the inner city found that the biggest attraction of the inner city lifestyle was 'the potential for diversity and cultural lifestyle enrichment' (Department of Planning and Urban Development (WA) 1993: 18). However, when asked 'what would stop you moving to the inner city?', the most dominant responses pertained to crime and violence (although it should be noted that this was a major concern for less than 20 per cent of those surveyed) (Department of Planning and Urban Development (WA) 1993: 4). New development tended in many cases to be security apartment blocks, or larger developments with their own private streets, and with little direct connection to the surrounding street environment. The flagship redevelopment of an old Council depot in Northbridge, for example, took up a city block, and comprised over 200 residential units. Designed by the same architects who had worked with Premier Court to produce *Perth – A City for People*, the units came equipped with remote-controlled security gates, intercoms and parking facilities. The development also included small, gated streets which acted as private internal byways for residents. According to the *West Australian's* Home of the Year awards, this complex, completed in 1996, was 'among the most spectacular developments to date' (*West Australian*, 26 July 1997: New Homes Liftout, 37). Spectacular it may have been, but it was somewhat removed from the European City ideal that planners had initially hoped to replicate.

These concerns about crime and safety were no doubt fuelled in part by the increasing amount of media attention devoted to juvenile crime and anti-social behaviour in the inner city throughout the 1990s. In 1994, the presence of young people (particularly Aboriginal young people) in Northbridge on Friday and Saturday nights replaced car theft as the new cause for concern. Northbridge is the main entertainment precinct in the Central Area, a place which the Central Area Policies Review Team hoped would continue 'to accept the newcomers, the back packers, low income earners and the adventurous, as well as the café society, as it has for so long' (Perth Central Area Policies Review Study Team 1993: ix). Nonetheless, the Northbridge Business Association, which had lobbied the PCC (successfully) to extend CCTV surveillance into Northbridge, complained to police about groups of young people loitering and disturbing their customers (*West Australian*, 25 May 1994: 45). The police 'JAG Team' (Juvenile Aid Group) used their powers under Child Welfare legislation to literally sweep the streets clean of those young people not accompanied by their parents. Policing operations with names like 'Operation Sweep' and 'Operation Family Values' were conducted with the intention of making 'the streets of the city and Northbridge safe for families' (*West Australian*, 26 May 1994: 9). The names of these operations are not insignificant – one suggesting that young people were dirt or rubbish to be swept away, the other suggesting that some normative ideal family was to be protected from this rubbish. On returning children to parents, police provided pamphlets and advice about parents' legal rights to use reasonable force to 'correct' their children's behaviour (*West Australian*, 26 May 1994: 9).

Police operations also targeted potentially vulnerable 'family members' who might have been led astray by so-called 'anti-social influences' – during Operation Family Values, for example, police removed young women from areas where Aborigines were thought to congregate (Blagg and Wilkie 1995: 41). Despite criticism from parents and youth groups, Police Minister Bob Wiese had no doubt about the success of the operations, telling the *West Australian* (7 May 1994: 28) 'Operation Sweep has been one of the most successful operations police have run while I've been Minister'. The Northbridge Business Association was also satisfied, claiming that the operations had contributed to a 35 per cent reduction in local crime. One of their number, however, worried that such operations would not provide a permanent solution to the problems caused by 'gangs' of young people:

> We can't just put them in a truck and ship them out to the bush, because they will come back and come back more angry, with more resentment (*West Australian*, 25 May 1994: 45).

Notwithstanding the concerns of such sensitive souls, WA police continued to conduct large-scale operations in the Central Area and

Northbridge. In 2003, for example, a similar police operation in North-bridge over three months netted over 450 children, almost 90 per cent of whom were Aboriginal (*West Australian*, 16 July 2003: 2).

The government also did all it could to cut off the flow of young people at its (suburban) source. In particular, it tightened regulation of public transport. In the 1980s and 1990s, suburban railway lines had been elec-trified and upgraded in anticipation of employment and retail growth in the Central Area, and the Central Railway Station was renovated and expanded to help it cope with the expanded capacity of the network (Alexander and Houghton 1994, 1995). It was not only suburban com-muters and shoppers who made the most of the expanded rail system. Young people from the suburbs also used the trains to access the Central Area. A street-based survey of over 400 young people questioned in the Central Area (which I will consider in more detail shortly) found that over 80 per cent of young people travelled to the city by public trans-port (City of Perth 1997: 22). This figure was as high as 90 per cent for a separate survey of Aboriginal young people conducted at around the same time (City of Perth 1997: 37). Not surprisingly, the rail network became the subject of increased public debate and calls for increased regulation. Parents were encouraged by one suburban mayor to prevent their teenage kids from catching trains into the city at night, because the trains attracted other young 'undesirables' (*West Australian*, 11 July 1997: 9). Once again, Aboriginal young people were singled out. Railways Commissioner Ross Drabble accused Aboriginal young people of being 'responsible for the most serious problems on Perth's trains' (*West Australian*, 6 March 1997: 7). The Aboriginal Legal Service counter-claimed that conflicts had only arisen because Aboriginal young people had been the victims of over-enthusiastic policing by transit police and railway security guards (*West Australian*, 6 March 1997: 7). While in Opposition, Labor accused the Coalition Government of failing to spend enough money on transit police. Labor Opposition leader Geoff Gallop said that passengers were still being 'terrorized by teenage louts', despite the introduction of $3 million worth of new security barriers and surveil-lance cameras throughout the rail network to 'exclude troublemakers from the Perth rail system' (*West Australian*, 21 July 1997: 7). As Premier, he committed an extra $7 million to electronic surveillance and transit police in the 2003 State Budget (Government of Western Australia 2003: 14).

At least, from the police's perspective, the trains provided a route *out* of the city centre as well as into it. For those young people that ran the gaunt-let of transit police and made it to the city and Northbridge, the JAG Team literally 'herded' anyone under the age of 18 onto the last train out of the city, at around 11 p.m.:

They call it the herding hour at Perth railway station . . . The big job for WA and transit police is getting hundreds of children on to trains and back to their homes before the public transport system closes down . . . It is when the attention turns from kids who might be drinking, drunk or high on spray paint, to anyone who looks under 18 and might be wandering city malls and Northbridge (*West Australian*, 21 March 1997: 10–11).

As in the case of other JAG operations, the concerns of local shopkeepers were important in prompting this police practice. They explicitly identified 'Aboriginal groups and street kids' as a problem (*West Australian*, 11 March 1997: 9; see also 10 March 1997: 3). City traders banded together to form Trader Safe to identify so-called 'hot spots' in the Central Area (*West Australian*, 12 March 1999: 36). Once again, it was 'families' who needed more protection in the Central Area. A Forrest Chase trader reminded a PCC Councillor 'we are both aware that Forrest Chase is being advertised and promoted as a family orientated area', while the manager of Forrest Chase wrote:

There certainly seems no point in spending heaps of money advertising for people to come to the city if when they get there the basic things of cleanliness and safety have not been addressed (*West Australian*, 10 March 1997: 3).

As a journalist who spent a night accompanying the JAG Team during herding hour noted of the young people being moved on:

These children are not the target of Perth City Council's *See You in the City* campaign – it is hard to find the smart-dressed parents with their college-cut kids and credit cards in tow (*West Australian*, 21 March 1997: 10).

Consolidating the Central Area as 'adult' space

If critics of exclusionary measures designed to secure the Central Area hoped that the re-election of the Labor Party to the State Government in 2001 would result in a significant change of approach, they were to be disappointed. Indeed, Labor Opposition Leader Geoff Gallop campaigned on the 'law and order' failures of the Coalition in both the 1996 and 2001 elections, and he maintained his Government's 'tough' stance upon taking office in 2001. The streets of the Central Area continue to be the site of a range of police interventions. In 2003, Gallop decided not to restrict government action to temporary policing operations to curb young people's use of the Central Area. As noted at the beginning of this chapter, a permanent night-time curfew is now in place for young people in Northbridge

Figure 6.3   Café life in the adult entertainment district: James St., Northbridge (photo: Kurt Iveson)

(see Figure 6.3). The *Young People in Northbridge Policy* instructs police to use their powers under Child Welfare legislation to remove unaccompanied children aged under 13 from the inner city 'adult entertainment precinct' of Northbridge[2] after dark, to remove children from 13–15 after 10 p.m., and to subject all others aged under 18 to a more 'hard line' approach. In its first year of the curfew's operation, 961 young people were removed from the streets of Northbridge. A staggering 88 per cent of those young people were Aboriginal.

Not surprisingly, the curfew has its critics, particularly the local Aboriginal Legal Service and the Youth Coalition of Western Australia. Both of these organizations have reasserted their position that the government and the PCC ought to offer youth-friendly spaces and services in the Central Area and Northbridge rather than seeking to exclude young people through confrontational policing operations (*West Australian*, 21 October 2003: 2). However, Premier Gallop has been unmoved by such criticisms:

> There are some leaders in our community who have expressed the point of view that we don't have any right to take these kids off the street. While you get some people saying that kids have a right to be there, that's sending a very, very bad signal out and it's making it harder for the Government to do what it should do on behalf of the community (*West Australian*, 21 October 2003: 2).

In particular, his Government has resisted any suggestion that young people ought to be welcomed into the Central Area:

> I say to the critics – start to consider the point of view of the victims – the people that live in this area, the people that work in this area, the people that visit this area (Gallop, quoted in *West Australian*, 21 October 2003: 2).

Once again, inner city living, employment and tourism are privileged as the 'proper' uses of the Central Area.

The curfew is not the only measure introduced by Gallop's government in an effort to keep unaccompanied young people out of Perth's Central Area. It has also sought to implement measures which will keep these young people in their 'proper' place – the suburbs and the family home. The lack of recreation facilities in the suburbs has been identified as one potential 'cause' of the Central Area's 'problem'. The Department of Community Development was said to be 'working with youth agencies in the suburbs to ensure young people were aware of safe activities and entertainment in their local areas' (*West Australian*, 9 January 2004: 5). The State Government has also identified families as an object of renewed govern-mental concern, with 'family dysfunction' explicitly linked to the problems of the Central Area. Kids, according to Gallop and his supporters, ought to be at home with their families rather than on the streets of the city centre. There could only be youths on the streets of the Central Area if some parents were failing to keep their children in the family home. So, while some families are considered ready to take their place (as shoppers, residents, etc.) in the Central Area, others have been condemned for contributing to the problems of the Central Area by failing to take 'respon-sibility' for their children. In 2004, Gallop announced plans to introduce new 'responsible parenting' orders and contracts. As the WA Government policy statement put it:

> Responsible parenting plays an important role in preventing children from skipping school or engaging in anti-social or criminal behaviour – and improving parenting skills has proven to be a successful early intervention for children or young people starting to show signs of that behaviour. The con-tracts and orders will be introduced to help parents create happier and more successful lives for their children, and in doing so, prevent their poor behav-iour from impacting on other people in the community (Office of Crime Prevention (Western Australia) 2004).

Under the plan:

> Mothers and fathers of truant, troublesome or criminal children would be forced to attend parenting classes to learn skills in communication, behav-

iour management, discipline and home-making. Those who refuse to attend would be punished (*West Australian*, 11 January 2004: 5).

Legislation for the parenting orders was introduced into Parliament in 2005, and at the time of writing the legislation is subject to legislative review. With the development of such strategies, the geography of interventions designed to secure the Central Area's 'public space' on behalf of 'the community' has expanded into the most apparently 'private' of institutions, the family.

City Challenges Pt 2

The range of interventions responding to challenges in the realms to design, marketing, social mix, security and family life did not necessarily fit together neatly into some coherent, coordinated package. On the contrary, they reflected a range of different agendas and the involvement of a range of agencies and interests. As such, tensions and contradictions began to emerge. In this section of the chapter, I want to move on to consider how these tensions and contradictions were identified and exploited by critics who challenged the model of 'public space' that was taking shape in Perth's Central Area. Of course, it should be no surprise that the governance of the Central Area has been contested. However, in this section, my aim is not simply to document the existence of contestation, but to assess the grounds on which alternative governance aims and mechanisms were proposed and implemented. As we shall see, in some instances these alternatives were offered as alternative solutions to a pre-defined 'problem', while in other instances it is possible to detect efforts to reshape and redefine the 'problem' that was to be solved by urban governance.

Re-visioning the city: from urban conflict to the 'harmonious city'

While we may be tempted to see only congruence between efforts to redesign and market the Central Area with efforts to secure the Central Area, these two sets of interventions also existed in tension. As we have seen, efforts to establish the Central Area as a terrain for a particular kind of 'public' – centred in particular on the retail and recreation activities of a normative family – resulted in calls for greater security and tighter spatial regulation to curb 'anti-social' behaviour. In making these calls, retailers were quick to point to those whom they considered to be perpetrators of such behaviour, and politicians routinely accused each other of failing to do enough to stop the breakdown of 'order' in the Central Area. However, the consequent publicity devoted to such issues threatened to discourage the very families and tourists that the city visionaries hoped to attract. City traders told the PCC they were:

> concerned that the City and Northbridge are perceived as unsafe, which is reinforced in media reports. As the statistics show, there has actually been a decline in most categories of reported and observed crime. It is not denying that some parts of Perth are at risk at some times for some crimes, however there is concern that inaccurate perceptions can and will deter people from coming into the City at certain times or even at all. Not only does it unnecessarily limit people's perceived options for shopping and entertainment, but it impacts economically on City businesses by reducing patronage (City of Perth 1999: 12).

City traders, of course, were among the sources for the very media reports contributing to negative perceptions of the inner city.

In order to address both the 'anti-social behaviour' *and* these 'inaccurate perceptions', both of which were thought to deter people from coming to the Central Area, WA Police conceived of their job as working on both 'reality' and 'perception' simultaneously. As one officer told the *West Australian*, attempts to deal with crime in the city must work across these two dimensions:

> It [crime] is more perception than it is actual. Certainly there's been marginal increases in certain types of offences and this campaign will reduce them to some degree. But the only way to turn people's perceptions around is to show them . . . that you're out there, you're highly visible, and that you are doing something (*West Australian*, 9 August 1997: 9).

In the attempt to address 'perception' as well as 'reality', then, a particular logic of policing and surveillance unfolded. And this logic was contradictory in some key respects. The yardstick against which the 'reality' of increases and reductions of crime was to be statistical reporting on numbers of thefts, assaults, etc. While statistics on such phenomena are of course imperfect, they at least offered some concrete measure of success for police and urban authorities. But how can one measure success in changing 'perceptions' of safety? According to the PCC and police, perceptions of safety are determined not just by levels of crime, but by levels of 'anti-social behaviour' which are not captured in crime statistics (1999: 7; see also Citysafe 1996: 4). The perpetrators of such anti-social behaviour can only be identified with reference to some set of norms. As Sercombe (1995: 90) has argued, ' "Potential offenders" do not declare themselves: they must be identified according to pre-existing schemas which shape the mind of a police officer'. Thus, it is not just the law but the normative boundaries of public space that have to be policed – in this case, the norms constructed in the city visions developed over the course of the decade. It is groups of young people unsupervised by their parents, particularly Aboriginal young people, who transgress these norms. If Aboriginal young people, and large groups more generally, are considered to be perpetrators of 'anti-social behaviour', then the very presence and visibility of these groups in public space is central to creating a perception that the city is unsafe. In response, police have established their own visibility in public space. They must be seen to be doing something to reassure members of 'the public' that they are in control, that the normative boundaries of the Central Area are being enforced. It is in this context that we can understand some of the police operational strategies discussed earlier in this chapter – police have conducted highly visible operations and foot patrols (and here, 'visibility' is achieved both through presence on the streets and through inviting journalists along to report on their activities), and surveillance cameras are installed in key locations not only in order to see what is happening but in order to be seen.

However, even this response produces unintended effects, further threatening the 'perceptions' of safety in the Central Area. The very fact that norms had to be patrolled through a highly visible police and surveillance presence meant that the Central Area continued to be represented as a dangerous and conflictual place. This contradiction made political space for another set of interventions, by a group of youth advocates who attempted to propose a new model of 'public space' premised on another vision of 'the City' as *harmonious* rather than conflictual. These advocates came to be involved in a range of policy-setting mechanisms designed to address the problems of 'anti-social behaviour' (see for example Citysafe 1996; City of Perth 1999). These planning mechanisms provided youth advocates (although rarely young people themselves) with opportunities to present different perspectives on youth crime and 'anti-social behaviour'. Of course, participation in such committees, and in wider debates about juvenile crime in the mass media, was constrained. Access was more likely to be granted to those who accepted the framing of the 'problem' to be solved by the committees, or discussed in the media, even if they disagreed about how it was to be 'solved'. As Sercombe (1995: 89) has noted in relation to participation in media debates, 'it is unthinkable that any commentator could say that they were not concerned about Aboriginal juvenile crime'.

Operating within this logic, in 1997 a group of youth advocates successfully lobbied the PCC to conduct a major study and policy development project concerning young people's use of public space in the Central Area. A survey of several hundred young people who used the inner city was conducted, alongside other qualitative consultations. Not surprisingly, it showed that young people liked coming to the city:

young people are mainly attracted to the city and Northbridge for social reasons and for the atmosphere it provides. This includes mainstream, ethnic, Aboriginal and at risk groups. It is therefore important to recognize that despite a lack of affordable things to do, the city has an atmosphere and special quality that will continue to attract young people (City of Perth 1997: 43).

Hopefully, the PCC planners and designers who had worked on *Perth – A City for People* initiatives were flattered – they had certainly succeeded in making the inner city the 'juxtaposition of suburban sameness'!

Over half of the young people surveyed complained of harassment by police, private security guards, transit police and shopkeepers (City of Perth 1997: 22). The authors argued that:

Attempts to exclude youth from these areas have generated anger, resentment, alienation and in some cases retaliation by young people. This harassment is most acutely directed at, and felt by, Aboriginal and 'at risk' youth (City of Perth 1997: 9).

Aside from this problem, young people had a range of other complaints about the inner city. Some identified other groups of young people as a problem – Aboriginal young people and those young people 'scabbing' for money were singled out as causes for concern by some young people, just as they were by retailers and police. 'Ironically,' the report noted, 'many young people group together for safety yet large groups can be intimidating to other people' (City of Perth 1997: 46). The relationship between young people and security agencies, then, is a complex one:

There is a paradox in that many young people see the need for more security, yet perceive police and security guards as 'against rather than for youth' (City of Perth 1997: 10).

In the face of these tensions in the Central Area, the report argued that 'inclusionary' policies, rather than harassment from police and security guards, would have more chance of creating what they called a 'harmonious city'. Not only would a harmonious city be of benefit to young people, but it would also help the PCC, the State Government and other city visionaries to address the negative perceptions (and reality?) of the Central Area as a dangerous, conflictual urban terrain. The authors made a series of recommendations to address the lack of late-night public transport, a lack of basic facilities like public toilets (the absence of which forced young people to seek entry to commercial premises), and a lack of low-cost recreational opportunities. Fifty-six per cent of all young people (and over 75 per cent of Aboriginal young people) surveyed came to the city with less than $20 (City of Perth 1997: 22, 37). The report further recommended that any

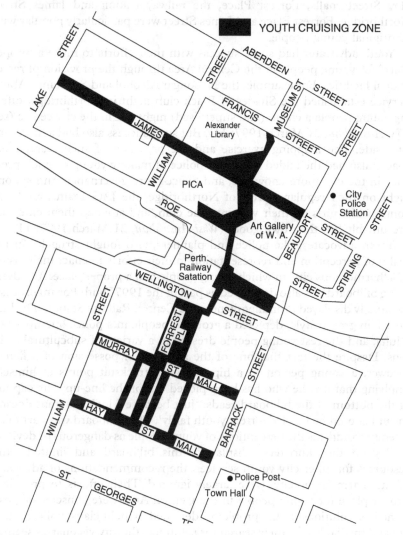

Figure 6.4   Map of 'Youth Cruising Zone' (source: City of Perth (1997) *Youth Forum: Report of Findings*)

new facilities for young people be located within young people's existing 'cruising zone'. A mapping exercise was conducted as part of the consultations, which was used to generate a pictorial representation of where young people 'cruised' in the inner city (City of Perth 1997: 8). This cruising zone fits very neatly onto a map of the CCTV network (see Figure 6.4). According to the survey, young people tended to congregate in Murray and

Hay Street malls, Forrest Place, the railway station and James Street Northbridge. Forrest Place and James Street were particularly popular with Aboriginal young people.

Youth advocates had some success with their efforts to make a 'proper place' for young people in the Central Area through the provision of recreational facilities. For example, the Nyoongar Alcohol and Substance Abuse Service established the Shades of Black club night in Northbridge, offering young people a cheap (and supervised) night out in the city centre (see *West Australian*, 21 March 1997: 11). But such success also had a more sinister side. The mapping exercise and the provision of such recreational spaces also had the added bonus for police of making young people's presence in the city more knowable, and hence more governable. And so, one night on their regular patrols of Northbridge, the JAG Team found two young white girls on their way into the club, and stopped them entering, presumably 'for their own good' (*West Australian*, 21 March 1997: 11).

Youth advocates, like police and planners, also fought struggles in the realm of perception and representation in their efforts to enact their vision of a 'harmonious city' in which young people had a proper place. An advertising billboard was also developed as part of the 1997 Youth Forum process, eventually displayed prominently near the Central Railway Station. The billboard, in graffiti style, depicted a group of people in a police line-up – two adults, and the rest young people dressed in a variety of subcultural fashions. Despite the fact that one of the adults is in possession of a flamethrower, a young person in a hip hop-style tracksuit points at himself, implying that it is he who has been picked out of the line-up. The caption at the bottom of the billboard reads: 'Just because I look different doesn't mean I should be hassled – treat youth fairly'. This billboard was part of an attempt to address the perceptions of young people as dangerous or deviant.

Beyond the short-term display of this billboard and limited extra resources for inner city youth services, the recommendations of advocates of the 'harmonious city' were largely ignored. Their efforts to provide a proper place for young people in the Central Area were consciously positioned as 'solutions' to the problems identified by politicians, police, retailers and media. But in their search for solutions, the city visionaries seemed unable to imagine a place for unaccompanied young people in the Central Area, and worried that any initiative which made the Central Area more attractive for young people would only exacerbate their problems.

The Nyoongar Patrol

As we have seen, throughout the 1990s, the behaviour of Aboriginal young people was explicitly identified as a problem by a range of actors involved in the governance of the Central Area's new public spaces. Premier Geoff Gallop's eagerness to establish Aboriginal communities' responsibility for the behaviour of Aboriginal young people could be characterized as an example of a discourse of 'responsibilization' that is typical of neo-liberal forms of social control (Rose 2000). While he argued that his Government needed to recognize and promote Aboriginal rights, he simultaneously asserted that 'unless there is a willingness to accept responsibility and make changes within Aboriginal communities, progress will not be possible' (Gallop 2003b: 3). To be sure, Gallop's approach was not restricted to 'get tough' measures such as the curfew and parenting orders – as part of a wider attempt to address anti-social behaviour and indigenous disadvantage, his Government also rolled out a series of early intervention and family assistance packages designed to improve parenting skills in Aboriginal families (Gallop 2003b; Western Australian Department of Indigenous Affairs 2005). As Mitchell Dean (2002) has observed, such approaches to government thereby combine *facilitation* and *authoritarianism* in order to achieve good order and social control.

Of course, the Gallop Government's approach to dealing with Aboriginal families and young people has had its opponents. In particular, both the Aboriginal Legal Service and the Youth Affairs Council of Western Australia have been highly critical of the Northbridge curfew and the policing of indigenous young people more generally (see for example Koch 2003; Youth Affairs Council of Western Australia 2003). Nonetheless, there has also been some degree of convergence between the neo-liberal discourse of 'responsibility' and 'self-management' and Aboriginal political discourses of 'self-determination'. Like neo-liberals, Aboriginal activists have been highly critical of past practices of liberal welfarism, if for quite different reasons. Demands for self-determination have emerged as critical responses to past and present colonial injustices, and these demands have extended into the justice arena (Blagg 2005: 14–15).

The Nyoongar Patrol in Perth is an example of one indigenous initiative informed by notions of self-determination that may at first appear to fit quite neatly within the neo-liberal approach to social control promoted by the Gallop Government. The Nyoongar Patrol is a nightly street patrol conducted by teams of local Aboriginal people, involving representatives from a number of Perth-based families and communities. Members of the patrol work the streets of Perth city centre to offer support, advocacy and service referrals for Aboriginal people they encounter. It is one of 20 Aboriginal Community Patrols in Western Australia which work to 'divert indigenous people from a diversity of potential hazards and conflicts' (Blagg and Valuri 2004: 207). The Patrol is supported by both the State Government and Perth City Council, and liaises closely with the Police Juvenile Aid Group during its operations. Like visions of the 'harmonious city' discussed above, the Nyoongar Patrol seems to provide an alternative solution to the 'challenges' (such as anti-social behaviour by Aboriginal young people) identified by the city visionaries.

Indeed, the city visionaries have attempted to mobilize the Nyoongar Patrol as part of their wider response to anti-social behaviour. Like other Aboriginal Community Patrols, the Nyoongar Patrol has been subject to pressure to fall into line with existing efforts to secure law and order in the Central Area:

> The policing of Aboriginal people has frequently been about close surveil-lance in, and outright denial of access to, public space. It is not inconceiv-able, therefore, that the same 'exclusionary logic' will shape the expectations of police, local government, business, resident and other partners in local security networks where initiatives affecting Aboriginal people are concerned (Blagg and Valuri 2004: 210).

Some public statements by people involved in the Patrol seem to accord with existing law and order strategies. For example, during the debates about the accessibility of the Central Area to suburban young people, the Chief Executive of the Nyoongar Patrol complained that these young people were coming to the city centre because there was simply nothing for them to do in the suburbs:

> There hasn't been any research to say that if you have activities in the suburbs it would stop kids from coming into Northbridge, but there's nothing wrong with trying (quoted on *ABC News Online*, October 20 2003).[3]

In such comments, the Patrol's Chief Executive seems to agree with Gallop's and others' framing of Aboriginal young people's presence on the streets at night as a 'problem' in need of 'solution'.

However, it would be a mistake to see the Nyoongar Patrol as operating wholly within the logic of social control that has been pushed by the coali-tion of interests involved in the reshaping of Perth's Central Area. Research conducted by Harry Blagg and Giulietta Veluri (2004: 213) suggests that while a range of city visionaries have sought to mobilize the Nyoongar Patrol to serve their own purposes, the Patrol has been vigilant in main-taining its own distinct sense of purpose:

> the Northbridge Retails Association, the Perth City Council and the state's peak crime prevention body (Safer WA) had . . . ambitions for Nyoongar – solving the 'problem' of Aboriginal young by removing them from the streets and, hopefully, luring nervous punters back to enjoy uninhibited *al fresco* dining. Nyoongar has consistently refused to fulfil this role and the disap-pointed retailers concluded that it had 'failed' (at a task it never set itself).

Nyoongar sees itself as existing to do more than simply meet the interests of the Northbridge business community. Business people want Nyoongar to be publicly funded security officers and execute a night time curfew for Aboriginal youth in Northbridge, while Nyoongar themselves see their role as proving a 'support service'.

The 'support' provided by the Patrol is framed as a response to the *vulnerability* of Aboriginal young people and rough sleepers in the Central Area, who were at risk of violence and exploitation through their street presence. In other words, the street presence of Aboriginal young people is not a problem *per se* – rather, the 'problem' which the Patrol seeks to address is the potential for harm associated with that presence (including the potential of arrest by police). So, rather than seeing the work performed by the Patrol as simply fitting in with existing security agendas, we might instead see the Patrol's existence as one example of how the very logic of social control established by the Gallop State Government is open to a range of interpretations and mobilizations for a variety of purposes. Of course, this 'openness' cannot be taken for granted – the Patrol has struggled to operate its own agenda in a difficult and frequently hostile context, and this context has proved fatal for other agencies working with Aboriginal young people.

The life and death of *GiBBER* magazine

Unlike the Nyoongar Patrol, which managed to secure its survival in a hostile context by exploiting the common ground of neo-liberal responsibility and indigenous self-determination, *GiBBER* magazine was killed off for daring to openly contest dominant visions of the city. *GiBBER* was a short-lived Perth-based magazine for young people. Its production was coordinated by two youth workers, who sought contributions in the form of writing or artwork from young people between the ages of 12 and 18 'who don't usually have their say. This could mean anything from not being at school to being homeless or in Longmore [juvenile detention centre]'.

The following poem, written by a young graffiti writer from the C.L.F. crew, appeared in the magazine *GiBBER* in 1995.

NO FUN. NO HOPE. DON,T BELONG[4]

C.L.F. Criminal liberation Front
C.L.F where chaos loving Freaks
C.L.F. Coz lifes Fucked
C.L.F. that's why we paint your streets
UP CHUCK, dont push your luck
        dontgive afuck
        when I run amuck.

> . . .
> Acrew ofkids and social misfits
> wreckin your sterilitys how we get our kicks
> A crew of chaotic, spray can junkies
> Bet you never seen nothin so funky.
>
> You canstuff your perfect, clean society
> Our paints our soals, screamin to be free]
> smash your order and power controll
> coz lifes to be lived on a roll.
>
> Every ones a criminal in the eyes of the state
> controlled and programmed is our fate
> Grafittys a outlet for our misguided hate
> Coz all chance for change is to late.
> The walls a canvass for my torched soal
> Communitys rejection of me is takin its toll
> Fuck you and your stale world
> Hope I fucked your day with the brick I herled.
> We blow our load on your clean world
> We paint over your clean wall
> Economic warfare evens the score
> Coz in ya wallets the only place you ever feel sore . . .
>
> VOID C.L.F. (in *GiBBER*, no7, 1995)[5]

It was *GiBBER's* policy to provide a space in which incarcerated young people could express and represent themselves. The workers engaged in outreach to alleyways, leisure centres, detention centres, and any other spaces where they thought they could find their target contributors and audience. *GiBBER* published all sorts of material – from angry poetry like VOID's to stories of pets and love and friends lost to the street, to artwork and discussions of drugs and alcohol use. Reading through the old editions of *GiBBER* gives one a remarkable window onto the ideas, perspectives and desires of some young people that is too rarely seen. Participation in *GiBBER* was not premised on adopting any particular perspective, least of all one which fit within visions of 'urban renaissance' or the 'harmonious city' – as VOID's poem ably attests. Predictably, this did not make the magazine popular with government agencies. The editorial of the ninth issue of *GiBBER* began by telling readers that the editors had been contacted by the Juvenile Justice Department of WA regarding the magazine's content:

The Department apparently believes that the content – content which is the expression of issues and concerns relevant to young people – needs to be censored or 'toned down'. Continued access to WA's Detention Centres may be revoked unless GiBBER complies with – to date – verbal ultimatum.

The letter from the Manager of Education at the Ministry of Justice to *GiBBER* read in part:

> as a government body entrusted with the care and welfare of the young people in their charge, Juvenile Justice is open to serious criticism should it continue contributing to and receiving the magazine. While Gibber's print policy of encouraging alienated youth to express publicly their innermost thoughts and emotions has merit, it is a teacher's responsibility to manage that expression in terms that are community appropriate and acceptable (quoted in Jefferys nd).

The ninth issue of *GiBBER* was the last. The exclusion of some young people from the 'the city' on behalf of 'the community' is pursued across a range of spaces – from the 'public spaces' of the city streets to 'public spaces' of media such as *GiBBER* where alternative perspectives on 'the city' might be articulated and circulated.

Conclusion: 'Public Space' and Neo-liberal Urban Governance in Perth

Place-marketing, gentrification, surveillance cameras, police sweeps, curfews, parental orders – of course, none of these measures is unique to Perth. The remaking of Perth's city centre by a loose coalition of state and private sector actors sought to attract a quite particular set of uses and users to this space while securing it against a range of uses and users that came to be defined as a threat. This combination of facilitation and authoritarianism is characteristic of neo-liberal forms of governance and social control (Dean 2002). However, this chapter demonstrates the long route towards such measures. The design and regulation of 'public space' in Perth was *contingent* on the efforts of a range of actors, who sought to envision and enact the restructuring of the city centre as a privileged space of the 'capital city'. As such, the representation and reconstruction of 'public space' was organized through two decades of public debates about the needs of the city centre and how these needs should best be met. Certainly, ideas from elsewhere played an important role in these debates. For example, Gallop acknowledged his Government's debt to Blairite policies on anti-social behaviour, telling the *West Australian* that 'The Parental Responsibility Orders are based on evidence of what's going on in Britain' (18 January 2004: 5).[6] But such measures have only been debated and adopted to the extent that they are perceived to have utility in addressing the 'city challenges' which arose in Perth as the urban coalition sought to turn their vision into reality. Foucault's (1980: 159) understanding of the diffusion of liberal techniques and technologies of government could equally be applied to the story of neo-liberal public space 'reform' in Perth that I have presented here:

> These tactics were invented and organized from the starting points of local conditions and particular needs. They took shape in piecemeal fashion, prior to any class strategy designed to weld them into vast, coherent ensembles.

By approaching the production of 'public space' through an investigation of how the streets became on object of concern in debates about the city's needs, as well as focusing on action in the streets, we can make three concluding observations about urban restructuring in Perth that point towards the complex geographies of neo-liberal urban governance discourses and techniques. First, it would be a mistake to construe events in Perth as an attack on, or destruction of, 'public space'. It may be tempting to make such a characterization, given the imposition of surveillance and policing measures in particular. But Perth's public spaces were not previously 'public' in the sense of being equally accessible to all. For example, Aboriginal people's presence in the Central Area has been problematized previously by urban authorities – between 1927 and 1954, the inner city of Perth was a prohibited area for Aboriginal people without permits (Haebicc 1992: 18; Jacobs 1996: 108). But there is also a wider point to be made here about how we understand 'neo-liberal' urban governance. It is perhaps more useful to consider what has happened in Perth as the *production* of (a particular model of) public space, rather than its destruction. The various efforts of those involved in Perth's urban restructuring were in fact directed towards producing the Central Area itself as a political subject with particular needs and interests, and the measures they imposed have been justified on the grounds of protecting the city/subject from threat. This approach accords with Larner (2003: 511), who notes that too often:

> geographical discussions of neoliberalism continue to focus on what we have lost. The possibility of thinking about neoliberalism as involving processes that *produce* spaces, states, and subjects in complex and multiple forms is downplayed.

Second, to the extent that the relative merits of 'exclusion' and 'inclusion' approaches to the creation of 'public space' were posed as alternatives, they were often posed as alternatives to a shared problem. So, for example, the problem of 'youth on the streets' could either be solved by keeping them out of the city altogether through surveillance and policing, or by allocating them a proper place in the city under the watchful eye of adults providing youth facilities in the 'harmonious city'. I do not mean to suggest that these approaches are the same, but I do want to suggest that they may well constitute different responses to the same concern (see also Iveson 2006b; Cohen 1985). Indeed, a wide range of measures have been posed as 'solutions' to the problem of youth on the streets – from action seeking to physically remove them from the streets, to action seeking to keep them in the suburbs by securing the public transport system or providing recreational facilities, to action seeking to bolster parents' capacity to discipline their children, etc. In fact, it could be said that the coalition pushing a particular vision of the city is held together more by a shared understanding of 'problems' and 'challenges' for governmental intervention than by a commitment to shared solutions. As such, the processes through which shared concerns are established and debated assume particular importance. We must not only ask: how accessible are 'public spaces'? We must also ask: how accessible are the public fora through which the identities of urban places are shaped and debated? This is an issue I discuss directly in the next chapter.

My third conclusion concerns the accessibility of these public fora and debates about the identity of Perth's Central Area. It would appear that certain groups (young people, and Nyoongar young people in particular) have been positioned as outside of 'the city' as a 'public' not only physically, but also procedurally – in terms I borrowed from Rancière (1999: 23) in Chapter 2, they have been denied 'symbolic enrolment in the city' as well as being denied access to its physical 'public spaces'. As Sercombe (1995) noted, it has been virtually impossible to gain access to the debates about Perth's inner city without some acknowledgement that young people are a problem, and even a threat, to its community. This has put young people and their advocates in a difficult position. Not all have chosen to comply fully with the underlying structures of understanding which constitute the unaccompanied presence of young people on the streets as a problem. The coordinators of GiBBER magazine chose to prioritize the problems identified by marginalized young people over the concerns of the city visionaries, with depressingly predictable results. For those who have been prepared to frame their alternatives as new solutions to problems defined by city visionaries, there have been mixed results. Advocates of the 'harmonious city' crafted a well-reasoned alternative solution to the problems encountered by city visionaries, only to be largely ignored.

Perhaps more promising is the experience of the Nyoongar Patrol. The aims of the Patrol do coincide with some elements of the dominant agenda for Perth, but they also exceed it. If we analysed the Patrol by starting with a pre-determined model of 'neo-liberal urban governance' and then looking for it, we might only see these apparent alignments. But if we focus on the contingency of neo-liberalism, as I have done here, other dimensions of the Patrol's work come into view. In particular, we can see 'self-governance' not only as a tool of capital and the state but also as a tool of struggle against post-colonial domination. In this context, geographers and others interested in the political-economy of urban restructuring would do well to heed Blagg and Veluri's (2004: 205) warning to criminologists fixated on the proliferation of neo-liberal forms of social control:

> Criminologists may find that some of their conceptual instruments, pre-packaged in the Euro-American criminological tool-kit, may be less useful in interrogating the meanings of these Indigenous initiatives in their totality.

Here, perhaps, is where some hope lies for alternatives to Perth's current path of restructuring. Not in a radical rejection of 'exclusion' in the name of an 'anything goes' inclusion, but rather in the crafting of institutions which might produce a different vision of the problems to be solved by urban governance. As Pat O'Malley (1998: 170) writes of indigenous forms of governance more generally:

> If projects for Aboriginal self-determination (or any other process of in-corporating indigenous governances) can result in the inscription of quite alien elements into liberal arrangements, the question that those concerned with the future of liberal governance may begin to ask is: how far can liberalism reach without becoming prone to internal fissions and substantial contradictions?

In the example of the Nyoongar Patrol, we have an example of how such fissions and contradictions might not only be identified, but how they might also be exploited in the creation of alternative governance agendas and institutions.

Chapter Seven

Justifying Exclusion: Keeping Men out of the Ladies' Baths, Sydney

The kind of street-based exclusion we have just examined in Perth has driven much of the recent literature on urban 'public space'. A good deal of this work is informed by, and supportive of, struggles to identify and contest the exclusionary characteristics of actually existing urban spaces. As I observed in Chapter 1, the notion that a truly democratic 'public space' ought to be 'open to all' tends to underpin these critiques of exclusion. The implication here is that alternatives to contemporary (often neo-liberal) forms of exclusion are some form of inclusionary or open public space. In the previous chapter, I pointed towards the limits of this notion, showing how both exclusion and inclusion on the streets can be mobilized in the service of social control. Exclusion from 'the city' takes a range of forms, and even when people are 'included' in the 'public spaces' of the street, they may nonetheless be denied access to the public discussions and debates through which 'the city' takes shape as a collective subject. In this chapter, I want to push this argument further, by arguing that alternatives to neo-liberal exclusions may involve *different forms of exclusion*, not simply 'inclusion'. My argument is that some forms of exclusion might actually be justified in order to sustain different forms of public address and sociability in the city.

The chapter focuses on a legal and political dispute over the exclusion of men from a public swimming pool in Sydney. It seems to me that this swimming pool presents us with an example of how spatial exclusion serves to facilitate the development and practice of counterpublic forms of stranger sociability which are consistent with radical democratic understandings of a heterogeneous public sphere. Typically, when advocates of a heterogeneous public sphere turn their attention to how the spaces of city might contribute to democracy, they focus on the capacity of 'public spaces' to sustain the mingling of diverse publics in a shared space. Consider, for example, Iris Marion Young's (2000: 213–14) discussion of 'embodied public space' and its relationship to a heterogeneous public sphere characterized by the interaction of multiple publics:

> A public sphere may be enacted partly through print and electronic media, and to that extent does not require open physical spaces. To the extent that physical public space shrinks, however, or to the extent that many citizens withdraw from embodied public space, open communicative democracy is in danger ... The physical open spaces of public streets, squares, plazas and parks are what I have in mind with the term embodied public space. These are large spaces where many people can be present together, seeing, being seen, exposed to one another. In them one may encounter anyone who lives in the city or region as well as outsiders passing through. They importantly contribute to democratic inclusion because they bring differently positioned strangers into one another's presence; they make concrete the fact that people of differing tastes, interests, needs, and life circumstances dwell together in a city or region.

Such interaction may be one kind of antidote to the kinds of exclusion established in Perth. But where precisely are the spaces in which people might develop the 'differing tastes, interests, needs and life circumstances' which constitute the heterogeneous public sphere? The very production of multiple publics involves the making of some spaces through which participants in counterpublics might establish alternative norms of dialogue and co-presence which are not subject to interventions from more powerful groups (Fraser 1992: 123; Iveson 1998). This process might take place anywhere, but it must take place somewhere. As we have seen in previous chapters, some people might find ways to sustain different forms of (counter)public address and sociability in sites which are shared with others. However, in other instances, participants in counterpublics might mobilize 'spaces of their own' in order to conduct dialogue over needs, values, interests and identities. They may turn apparently 'private spaces' such as a living room or a kitchen into a 'public space' by coming together to plan a rally, produce a pamphlet, or share experiences (see for example Fincher and Panelli 2001). They might also make claims for exclusive access to 'embodied public spaces' as spaces of withdrawal from the wider public, either at particular times or even permanently. Here, counterpublic norms of dialogue and co-presence are formed through exclusionary occupations of particular urban spaces. Of course, this raises very important questions about how such exclusions might be enacted and justified.

While the exclusions discussed in previous chapters are generally secured and normalized with reference to private property rights and pre-political visions of shared 'community values', the struggle over McIvers is particularly interesting because the women who campaigned to exclude men from McIvers were forced to make a public case for their legal and political rights to exclude men from an apparently 'public' place. Here, as we shall see, the concept of a universally accessible 'public space' was mobilized by men who wanted access to McIvers, and opposed by women who sought to protect a space which enabled the construction of counterpublic norms of being together with strangers (albeit strangers of a particular gender). In this dispute perhaps more than any other in this book, the limitations of a liberal model of universally accessible 'public space' for a critical, non-liberal theory of publics and the city are laid bare. The chapter proceeds by outlining the history of the battle for McIvers, before offering an extended analysis of the meanings of publicness mobilized by parties to the dispute in order to make this case.

The Dispute

A thirty-minute bus ride from Sydney's Central Business District, through some of Australia's highest-density residential neighbourhoods, takes you to Coogee Beach. Coogee is one of Sydney's busiest ocean beaches – popular with local residents, other Sydneysiders and tourists all year round, its nearby roads are lined with cafés, restaurants, apartment blocks, backpackers' accommodation, bars and other leisure and commercial facilities. Nestled at the foot of coastal cliffs a few hundred metres south of the beach is McIvers Baths. It's a gorgeous spot, carved out of the rocks over one hundred years ago to provide bathers with some protection from the ocean swell and prying eyes. In contrast to the crowded sand, surf and streets at Coogee Beach, McIvers offers a far more tranquil experience, with a pool for gentle lap swimming as well as a small patch of grass and large coastal rocks for those who want to soak up some sun. It is one of a number of surviving ocean pools constructed in the early years of the twentieth century along Australia's east coast.

For all of its history, men have been excluded from McIvers. While McIvers was not the only ocean pool in NSW to begin its life as a gender-segregated facility, the others had been gradually desegregated over the course of the twentieth century (McDermott 2005). Since 1922, the pool has been managed by the Randwick and Coogee Ladies Amateur Swimming Club (RCLSC), which asks swimmers to leave twenty cents in a bucket at the entrance to the pool to help with maintenance costs. McIvers regularly features in newspaper and magazine guides to swimming spots in Sydney. These articles typically include an anecdote or two about men who have had to be chased away by vigilant pool users. Lorna Mobbs, a member of the RCLSC, once told a Sunday newspaper, 'I don't blame a man for wanting to get into the women's baths. It must drive them mad'

(*Sun Herald*, 12 December 2000, 'Sunday Life': 13). Mobbs could joke about it then, but a few years before this was no laughing matter. The exclusion of men from McIvers became the subject of a lengthy legal and political dispute which threatened its women-only status.

The dispute began on 16 December 1992, when a male resident of Coogee complained to the NSW Anti-Discrimination Board of his exclusion from McIvers, alleging discrimination on the grounds of sex. The letter of complaint said in part:

> At Coogee Beach there is a seawater swimming pool fenced off with a sign at the entrance stating 'Ladies' and Children's Pool'. Is this permissible? If it is permissible, why? Should it not be permissible, what is necessary to have this sign removed? (*Wolk v Randwick City Council*, 1995)

Over the next two years, the complainant spent around $10,000 in his effort to open McIvers to men (*The Messenger*, 9 May 1995: 6).

The 1992 complaint was not the first challenge to the exclusion of men from McIvers. In 1946 a proposal to open the baths to men was rejected by the Randwick Council Works Committee (*Randwick and Coogee Weekly*, 10 October 1946). However, the 1992 letter was the first formal complaint about McIvers made under the *Anti-Discrimination Act (NSW) 1977 (ADA)*. Some had feared that a complaint under the ADA's sex discrimination provisions would happen eventually. A writer for Sydney magazine *Lesbians on the Loose* related the following experience of trying to get information about the pool for an article early in 1992:

> Part of the survival of the pool has been that the women who run it let very little information out. Try to find the pool if you don't know where it is – there are no signs until you are right on top of it . . . When I first approached the woman in the club room at Coogee pool for information, she slapped me lightly on both cheeks and said 'Sorry, Lesley, we don't want any publicity' . . . They are scared, and rightly so, that if more people (read men) know about the pool they will try to take it over or close it down (*Lesbians on the Loose*, April 1992: 14–15).

Such fears about publicity should not have been necessary. The activities of single-sex sports clubs or volunteer associations such as the RCLSC are not covered by the *ADA*. However, the RCLSC only managed McIvers – Randwick Municipal Council (hereafter 'The Council') leased the site from the Crown, and was therefore considered 'the owner' of McIvers for the purposes of the complaint. In a letter to a member of the RCLSC, President of the Anti-Discrimination Board Chris Puplick noted that the Council's failure to prevent a challenge to McIvers, either by transferring the lease to the Club or by applying for an exemption under the *ADA*, was 'regrettable' (Puplick 1995).

Initially, the Council opposed the attempt to open McIvers to men. When the complainant and the Council failed to reach an agreement, the Anti-Discrimination Board established a conciliation process during 1994, pursuant to the *ADA*. The complainant suggested that the pool be open to both men and women in the mornings, and reserved for women and children in the afternoon. This proposal was in line with his own desire to have a place to swim in mornings when rough seas might prevent him swimming elsewhere, but was unsatisfactory to both the Council and the RCLSC (which had been admitted as a party to proceedings) (*Sydney Morning Herald*, 17 January 1995: 3; *Weekly Southern Courier*, 17 January 1995: 10). They continued to oppose the introduction of any mixed bathing at McIvers. The complaint attracted a small amount of publicity at the time, a lone article in the *Sydney Morning Herald* (11 February 1994: 4) outlining the nature of the dispute.

With the failure of conciliation, the dispute was listed for arbitration by the NSW Equal Opportunity Tribunal (EOT). Up to this point, the RCLSC's efforts to 'save' McIvers had focused on contesting the complaint in conciliation proceedings. However, with the right to maintain McIvers exclusively for women and children now being arbitrated by the EOT, they mobilized a wider campaign in defence of the pool which targeted both the local press and mass media such as the *Sydney Morning Herald*. The RCLSC and their supporters argued that there were plenty of other places for men to swim. Tides only prevented swimming in other ocean pools for around twelve days each year, and in any case, McIvers was too short and shallow for vigorous lap swimming. It was also argued that a range of women with special needs who made use of McIvers would not be able to swim there if men were present.

The Council sought legal advice, and was advised that the chances of success in an arbitrated decision were slim (Randwick Municipal Council 1995a: Q4.1). Newspapers on Tuesday 17 January reported that in light of this advice, Randwick Mayor Chris Bastic had decided not to contest the complaint at the EOT. The costs of a legal defence of McIvers were too high, he said, given the slim chances of success (*Sydney Morning Herald*, 17 January 1995: 3; *Weekly Southern Courier*, 17 January 1995: 10). Mayor Bastic's withdrawal of support drew a loud and angry response from the RCLSC and its supporters. On Wednesday 18 January, the *Sydney Morning Herald* reported that Bastic had changed his mind, owing to the 'overwhelming support' for retaining McIvers exclusively for the use of women. On the same day, its editorial focused on the dispute, urging the complainant and others like him to 'leave these women alone'. That night, the meeting of the Council was crowded with supporters of McIvers. Jane

Campion, the film-maker riding high on the recent success of the movie *The Piano*, had been recruited to further elevate the profile of the campaign to save McIvers. Campion and Vivienne McCredie, Secretary of the Club, both addressed the meeting. The Council unanimously passed a motion:

> That McIvers Ladies Pool at Coogee be retained for the exclusive use of women and children under twelve years and that Council take all necessary steps to support the present use of the pool, including proceedings before the Equal Opportunity Tribunal (Randwick Municipal Council 1995a: 2).

The *Sydney Morning Herald* published a steady stream of letters about the McIvers dispute in the two weeks following its editorial. There were plenty of letters in support of opening McIvers to men, although the Letters Editor commented that more had supported the fight to keep the pool for women and children only (*Sydney Morning Herald*, 23 January 1995: 10). In any case, support for opening the pool to men was restricted to individual letter writers – no organizations or institutions formally supported the complainant, either directly or indirectly in the wider public sphere. Those opposed to allowing men access to McIvers, on the other hand, mobilized a wide range of women's organizations behind the cause, including the Women's Electoral Lobby and the Women Lawyers Association of NSW. The Kingsford Legal Centre, a community legal centre which specializes in anti-discrimination law, was engaged to act on behalf of the RCLSC.[1] Kingsford Legal Centre's lawyers argued that the *ADA* was able to accommodate women-only spaces like McIvers. At the height of the public interest in the dispute, Kingsford Legal Centre Director Simon Rice wrote an opinion piece for the *Sydney Morning Herald* (31 January 1995: 13), outlining the case in favour of keeping men out of McIvers. He argued that anti-discrimination law did not necessarily imply equal treatment, but could also be used to 'protect difference' in cases where this did not result in arbitrary and unfair discrimination against individuals.

In the meantime, a meeting of the Council, the RCLSC, and their legal representatives decided to approach the Minister for Local Government, Liberal Party MP Ted Pickering, to secure certification of McIvers as a 'special needs activity' under section 126A of the *ADA*. Section 126A allows the Minister to exempt Special Needs Programmes and Activities from the *ADA*, where s/he is:

> satisfied that its purpose or primary purpose is the promotion of access, for members of a group of persons affected by any form of unlawful discrimination, . . . to facilities, services or opportunities to meet their special needs or the promotion of equal or improved access for them to facilities, services and opportunities.

The Council applied to the Minister for this exemption, with a supporting application lodged by Kingsford Legal Centre on behalf of the RCLSC. The applications mobilized a range of arguments which were both locally specific and more general. There were three main arguments pertaining specifically to McIvers and its relationship to the Coogee/Randwick area. First, the applications focused on the long history of the pool as a space for women, and noted the important role played by the RCLSC in maintaining McIvers and providing other volunteer services. It was put to the Minister that this arrangement would be sure to collapse if men were admitted. Second, it was noted that other coastal pools in the area, including Wylies Baths, were available for men. Third, the applications argued that particular groups of women would no longer be able to use McIvers if men were admitted. Kingsford Legal Centre's letter to the Minister included supporting documents from Randwick Girls' High School, two local women's refuges, and the local Women's Support Network. The Girls' High School held their swimming lessons at McIvers because, according to the school sports organizer:

> we have a high proportion of Muslim girls in our classes who cannot use any pool other than a girls and women-only pool. Before this, our Muslim girls were not able to choose swimming as an option during the sports class, because their religion forbids them from swimming with males . . . [M]any Muslim girls have learnt to swim in the ladies' pool who would never have done so without this arrangement (Kingsford Legal Centre 1995).

The letters from refuges noted the importance of retaining McIvers exclusively for women to allow their residents, many of whom had experienced domestic violence and rape, to continue to swim there. Kingsford Legal Centre had conducted a survey of women who used McIvers, and their submission also included comments from survey responses.

The applications for exemption located these locally specific arguments in favour of retaining McIvers exclusively for women in the context of the broader anti-discrimination objectives of the *ADA*. The Council argued that

> Opening up McIvers Baths to men would deter many, if not most of the current users, from going swimming at all. This would be detrimental to their physical and mental health. For as long as women suffer unlawful discrimination, particularly by way of sexual harassment, it is the belief of Council that they have a special need to a facility like McIvers Baths . . .
> By providing facilities which allow women to participate in social, sporting and health-related activities from which the presence of men would deter them, Randwick Council is bringing about equality. By ensuring women have exclusive use of McIvers Baths, Council is making its contribution to

bringing women up towards the level of enjoyment of such facilities which men have throughout the State (Randwick Municipal Council 1995b: 4, 9).

Similarly, Kingsford Legal Centre argued that the admission of men:

> would be a result which is contrary to the spirit of s126A of the Anti-Discrimination Act 1977, as the opportunity would be denied to women to access a facility which is sensitive to their special needs (Kingsford Legal Centre 1995: 4).

Fortuitously for those campaigning to keep McIvers for women only, Coogee was a marginal seat in the NSW State Election, to be held on 25 March 1995. While Opposition Leader Bob Carr was busy talking up a panic over juvenile gangs (see Chapter 5), his Labor colleague Ernie Page was supporting the campaign to keep McIvers for women only. The Coalition Government was targeted by a large-scale letter-writing campaign in support of the Council's application for certification under section 126A of the *ADA*. As the *Sydney Morning Herald* observed:

> Both Labor and Liberal candidates have sniffed the wind and are running hard for the baths to remain exclusively for the use of women. The question now is how this will be done.

Minister Pickering settled the question on 3 March, by granting McIvers special needs exemption from the *ADA* (Pickering 1995). This exemption protected McIvers from any action under the *ADA* in the future. Pickering's press release clearly demonstrates both the campaigners' effective mobilization on locally specific issues, and the importance of the forthcoming State and Council elections:

> In making this decision I have considered a range of factors, including the long history of McIvers Baths as a place reserved for women and children . . . I also wish to point out that the majority of representations on this sensitive issue, including those of Randwick Councillor Margaret Martin, have been in favour of retaining the traditional use of the baths (Pickering 1995).

Randwick Councillor Margaret Martin was a member of the Liberal Party, singled out here by her State coalition colleague to bolster her campaign to become Mayor of Randwick and to neutralize the significance of the issue for the State election. Martin's and Pickering's support for the pool was vital to nullify any political mileage that the Labor Party might have extracted from the dispute. The Ministerial exemption also recognized the appropriateness of special treatment in a wider sense, Pickering stating that:

I have taken into consideration the detailed submission by council on the issue of access for women with particular religious beliefs . . . Additionally, I am satisfied there are adequate nearby unrestricted swimming facilities (Pickering 1995).

While Pickering's exemption for McIvers was a major victory for the RCLSC and its supporters, it came too late to prevent the complainant having his day in court – his complaint of past discrimination was still to be heard by the EOT. While the decision of the Tribunal would not overturn the Ministerial exemption, it nonetheless provided both sides with further reason to justify their position. The Tribunal rejected the Council's claim that the initial letter about McIvers did not meet the technical criterion for a complaint. However, it accepted the Council's claim that as a matter of law, the complaint of discrimination related not to the 'provision of goods and services', but rather to 'access to a place'. Because of legislative oversight, discrimination on the grounds of gender and sexuality in relation to 'access to a place' is not covered by the *ADA*.[2] The complaint was therefore dismissed, with the decision stating that:

the material before the Tribunal does no more than indicate that the respondent, at the highest, restricted access to a place, namely, the swimming pool which is the subject of these proceedings, under its control (*Wolk v Randwick City Council*).

Extraordinarily, given the high public profile of the dispute, the Tribunal was also of the opinion that 'the allegations made in the letter of 16 December 1992 are frivolous or vexatious and should be dismissed on that basis'.

Equal or Special Treatment?

The outcome of this dispute over the women-only status of McIvers is remarkable. In Australia, while some public services for women have been successfully protected from complaints of discrimination (see for example the decision in *Proudfoot and Ors v Australian Capital Territory Board of Health*), attempts to establish women-only spaces in publicly funded leisure facilities in other state jurisdictions have had less success. Indeed, two previous attempts to establish limited women-only swimming periods at public pools in the state of Victoria, using similar arguments to those mounted in the campaign to protect McIvers, had failed (see the decision in *Pulis and Banfield v Moe City Council*, and Foulkes 1992).

Furthermore, the decision was controversial given simultaneous events in nearby Bondi Beach. In February 1995, the month before the Ministerial exemption for McIvers was granted and the EOT handed down its ruling, a decision had been made by the male-only Bondi Icebergs club to admit women for the first time in its sixty-five-year history.[3] The decision had followed small demonstrations by women outside the clubhouse, and the threat of a complaint to the Anti-Discrimination Board (*Sydney Morning Herald*, 28 February 1995: 3). The Bondi Icebergs managed the Bondi Baths (leased from Waverley Council), an ocean pool a few kilometres north of McIvers, as well as the clubhouse which included a sunbathing deck and a licensed club. Both the pool and the licensed club were accessible to men and women; only the sunbathing deck was restricted to Icebergs (male) members. In 1994, Waverley Council threatened to demolish the clubhouse, after an engineer's survey had found that the building was 'riddled with concrete cancer' (*Sydney Morning Herald*, 9 February 1994: 3). This would have meant an end to the Bondi Icebergs' management of the pool and clubhouse. Waverley Councillor Paul Pearce said that replacing the building with new premises for the Icebergs would be too expensive, and added 'I can't justify spending public money on a licensed bar for a men's only private club' (*Sydney Morning Herald*, 12 March 1994: 6).

Like the battle for McIvers, the threat to the Bondi Icebergs received substantial media attention – in this case, however, the dispute was covered generously in the press because of the fame of the Icebergs and their winter swimming rituals. Late in 1994, after the Icebergs and their supporters (including the women-only winter swimming club the Bondi Mermaids) mobilized opposition to the destruction of the clubhouse, Waverley Council withdrew its threat to demolish the clubhouse, and began to seek an alternative solution to the building's structural problems (*Sydney Morning Herald*, 10 January 1995: 10). In raising support for their cause, however, the Icebergs became more susceptible to claims of exclusion. Although most of the facilities managed by the Icebergs were accessible to women, only men were able to participate in this management through club membership. One of the main complaints of women who demonstrated outside the clubhouse was that they could not join the management committee involved in trying to improve the baths. The Icebergs' Treasurer recognized that although they had strong political support in the struggle with Waverley Council, these supporters 'wouldn't listen to us any more if we excluded women from all facilities'. A meeting of the Icebergs in February voted unanimously to make membership available to women (*Sydney Morning Herald*, 28 February 1995: 3).

Some people argued that the simultaneous campaigns to open the Icebergs to women and not to open McIvers to men were exemplary of a kind of hypocrisy in feminist politics. A letter attacking the *Sydney Morning Herald*'s editorial about McIvers remarked that in the campaign to open nearby Bondi Icebergs to women:

> the great principle was that a publicly funded facility should not exclude half the population, and the expounder of that principle was the woman mayor of Waverley. With absolute correctness, she said that if the men of the Icebergs wanted to go to the expense of building their own private pool, there would be no problem. Can your editorialists explain why all this does not apply to women in the Municipality of Randwick (*Sydney Morning Herald*, 21 January 1995)?

There were significant differences between events at Bondi and Coogee. While the dispute at Bondi related mostly to involvement in the management of a substantial community resource, the battle for McIvers was a battle over access. At Bondi, the Icebergs decided that it was in their collective interest to admit women to the club, and voted as such, while the RCLSC had made no such retreat. In fighting the complaint, the RCLSC put their case in terms of both tradition and social justice, while the arguments put as to why men at Bondi needed exclusive management of the pool and clubhouse relied on tradition.

Nonetheless, the contrast between events in Bondi and Coogee does serve to illustrate significant tensions over the nature of anti-discrimination principles and measures and their application to urban spaces and facilities. In Bondi, women campaigned for, and won, equal treatment. In Coogee, they campaigned for, and won, special treatment. In theory, anti-discrimination legislation in NSW and other jurisdictions in Australia recognizes the need for marginalized groups to be treated equally in some circumstances, and differently in others. The very possibility for an exemption was built into the *ADA* to allow for measures based on principles of affirmative action. While the decision of the EOT hardly constitutes a victory for affirmative action, based as it was on the distinction between access to a service and access to a place, the Ministerial exemption clearly recognized the merits of special treatment for women users of McIvers. But as the dispute over McIvers illustrates, the simultaneous provision of measures aimed at both equal treatment and the protection of difference has inevitably been contentious:

> Anti-discrimination legislation recognizes the need for both approaches through the mode based on lodgement of individual complaints, on the one hand, and the affirmative action mode, on the other hand. Nevertheless, the equal treatment approach, counterposed against a 'special treatment' approach, has inevitably generated a significant and irresolvable tension (Thornton 1990: 2).

Public or Private?

Central to these tensions is the complex positioning of 'special treatment' provisions (such as the exclusion of men from McIvers) with respect to contested understandings of the public–private distinction. For those who wanted McIvers opened to men, it could not be both a 'public space' or 'public facility' *and* exclusively for women and children – such exclusion could only be justified, according to one letter writer, if women 'wanted to go to the expense of *building their own private pool*' (*Sydney Morning Herald*, 21 January 1995, emphasis added). The meaning of publicness reflected in this position can be read as the product of two related logics. First, the complainant and his supporters asserted that what is public should be 'open to all', and that what is particular should be forced into privacy – a clear example of a liberal understanding of publicness (see Chapter 2). Second, there is a conflation of two distinct aspects of publicness – what have been referred to earlier as 'visibility' and 'collectivity' (Weintraub 1997). Simply put, it was argued that McIvers ought to be public (in the reaction of visibility) because it is collectively owned and provided by the state (on behalf of 'the public').

Those who opposed the complaint successfully refuted both of these logics. They argued that the ongoing exclusion of men from McIvers facilitated the practice of specific forms of public sociability and co-presence which were not possible in other public spaces, and that this particular exclusionary public space therefore both deserved and required state support in the form of legal protection and resource allocation. It is useful to consider this position in some depth, to see what light it may shed on the relationship between counterpublics and the conventional territorialization of the public/private distinction in cities.

The RCLSC and their supporters asserted that the absence of men from McIvers was fundamental to the protection of particular forms of sociability and interaction which are not possible in other public spaces that are universally accessible. Specifically, they identified McIvers as a place in which women and small children could escape an objectifying male gaze and male boisterousness. So, for example, surveys of pool users collected by Kingsford Legal Centre during the dispute reported that many women valued a women-only pool because it allowed them to swim at their own pace:

> Being elderly, of frail build with osteoporosis, this is the only salt water pool I can do my therapeutic laps in without being bumped by men.

> When doing lap swimming men have actually swum over the top of me forcing me underwater. They just take over and you can't get a proper swim.

Other respondents noted the absence of male 'perving' at McIvers, and the consequent difference between McIvers and other spaces where perving and/or harassment is the norm:

> I breathe a sigh of relief when I pop my 20 cents in that bucket – going into that beautiful, relaxed atmosphere knowing I can sun-bake and have a lovely swim without being hassled or perved at by men, or having my personal space invaded.

> With just women and small children it is quiet, peaceful and safe. There are not many places you can go without being subjected to the loud showing-off of young men. It is such a small pool that the presence of men would take up the entire space.

For those who supported the ongoing exclusion of men from McIvers, then, the pool was and is a space of withdrawal for women and children from the wider public sphere. This withdrawal accounts for some references to the 'privacy' afforded by the limited access to McIvers. Doris Hyde, for example, had once noted that McIvers was 'valued by women because it allows for privacy' (*Eastern Herald*, 5 December 1985: 17). But this 'privacy' for women should not be taken to fix the identity of McIvers wholly on one side of the public/private binary. Read from a non-liberal perspective, the exclusion of men from McIvers has facilitated the establishment of alternative norms of *public* sociability and co-presence among (particular, female) strangers. Randwick Council (1995b) made this point in their description of the purpose of McIvers as 'swimming and sunbathing, school and club activities, including learn-to-swim classes, for all female members of the public'. The ongoing exploration of identities, values, needs and interests through the space of McIvers is a public process of negotiation, discussion and debate among women who are strangers to each other. We can see at McIvers the formation of a kind of counterpublicness through the making of a public space of withdrawal from the wider public sphere.

The scope and diversity of the (counter)public dialogue over norms of co-presence at McIvers were reflected in the wide range of organizations and individuals involved in the campaign to oppose the complaint. Indeed, this diversity was fore-grounded in the campaign to save McIvers, and was crucial to the campaign's success. A coalition of women's groups was built, based on an acknowledgment of the different women who used the pool, for all sorts of different reasons, and in all sorts of different ways. A letter in support of the Club from the Women Lawyers' Association of NSW is typical, noting that the pool was important for:

> older women; women with disabilities; women who had undergone traumatic surgery . . . ; women from religious groups including many nuns; women from

diverse cultural backgrounds; women with small children, nursing mothers and pregnant women; women seeking sanctuary in the gentleness of our pool and the companionship of other women. Many women who are survivors of rape or domestic violence use the pool because it is safe and they do not have to continually worry about unwanted attention from some men.

The norms of sociability established at McIvers are quite particular to its status as a women-only place, even as they are nonetheless dynamic and themselves open to change. Certainly, McIvers is a space in which different expectations and practices are normalized at different times. These expectations and practices are not imposed by some private authority, but rather are negotiated and contested in a public dialogue involving the different women who use McIvers. So, for example, tradition remains an important feature of the pool's management, and the RCLSC continues to conduct free swimming lessons for children on Saturday mornings, and Thursday is still 'married ladies' day'. At these times, overt 'displays' of sexuality are discouraged, as is nude or topless bathing in the sunbathing shed or around the pool. But at other times, standards of behaviour and display are more relaxed, and different norms of interaction are established.

At one point during the dispute, the campaigners' public claims about the diversity of women who used McIvers were challenged. Un-named critics accused the pool of being dominated by a narrow range of women users. It was not a 'women's space', as proposed by the RCLSC and its supporters, but a 'lesbian lair' (*Weekly Southern Courier*, 15 February 1994: 1). In response, the Mayor said that he had 'no knowledge that the pool was used on occasions for lesbian activity', and RCLSC President Doris Hyde 'rejected suggestions that the pool was used by lesbians', saying that she 'had never seen anything untoward there' (*Weekly Southern Courier*, 13 September 1994: 4). It was never made clear in the press who was making homophobic allegations in the first place, nor why the use of the pool by lesbians should be a cause for concern. In 1993, before the dispute went public, Doris Hyde had been more candid in an article about swimming spots in Sydney, in which she said that she was aware of the pool's popularity with lesbians:

> they are the nicest girls and they come in and throw a couple of dollars in the till and say 'keep the change'. They speak nicely to me and they are the ones who'll put the fellows out (*Sydney Morning Herald*, 27 December 1993: 9).

Hyde went on to say that she was 'proud of that fact that such a diversity of women, including both the "oldies and youngies", continue to enjoy the pool'.

The existence of an active and diverse public dialogue among the women users of McIvers had its roots, of course, in the gendered segregation of recreational spaces which characterized urbanization in late nineteenth-century Australia (and elsewhere). At the time when McIvers was constructed, it was illegal for men and women to bathe together in public at all, and women were forced to withdraw to the 'public privacy' (Cooper, Law et al. 2000) of secluded spaces.[4] However, as writers like Elizabeth Wilson (Wilson 1991) have shown, women have historically used these marginal spaces to explore and construct new identities. By the time of the complaint, the long-term existence of McIvers as a women-only space had facilitated a set of alternative practices and spatial norms in which a diverse group of women were already engaged, and to which they had formed a commitment.

As well as establishing the existence of alternative and particular forms of sociability at McIvers, the campaigners successfully argued that these forms of sociability deserved ongoing state support in the form of legal protection and resource allocation. That is, they contested the logic that the state, acting in the collective or 'public' interest, should only provide spaces which are universally accessible in the liberal sense. State support for the exclusion of men was achieved in two ways – legally and politically. Legally, the RCLSC and their supporters argued that the exclusion of men was not contrary to the principles of anti-discrimination legislation. Politically, campaigners secured the support of both Randwick Council and the State Government through a sustained and wide engagement with a variety of publics. Debate ranged across a variety of institutions and media, from local and mass media to Council Chambers, marches such as International Women's Day, the State Parliament, and the courts. Indeed, campaigners expertly worked to expand the scale of the public discussion through which the identity of this public space was determined, arguing that access to McIvers was both a matter of public interest in the immediate locality and a matter of interest to women all over the State of New South Wales.

In reflecting on the form of public sociability that has been legally and politically secured at McIvers, then, it is important to keep in mind the conceptual distinction between different dimensions of publicness (as visibility and collectivity). We must be careful to distinguish between the openness of a particular site such as McIvers, and the openness of dialogue about that site and its norms of access (i.e. whether this dialogue is open to the participation of a range of publics). While men are excluded from the place, the campaigners engaged with a range of individuals, both men and women, across a variety of publics in sustaining the identity of McIvers. A relatively inclusionary political debate in the wider public sphere resulted in the democratic legitimation of the exclusive use of a particular public space.

So, while the exclusion of men could be seen to have produced a 'separatist' space, to label McIvers as such would be to confuse the conceptual distinction between 'embodied' public spaces and political public spheres. I agree with Deutsche (1996: 289) that 'the public sphere remains democratic only insofar as its exclusions are taken into account and open to contestation'. Importantly, however, this does not necessarily imply that all exclusions ought to be opposed – rather, it implies they ought to be determined democratically by debate among a variety of publics.

Conclusion: Separate Spaces Are Not (Necessarily) Separate Camps

Clearly, the arguments mobilized in the campaign to protect the women-only status of the pool were not necessarily consistent, or dictated by some coherent set of normative principles about the proper meaning of 'publicness'. Indeed, the existing arrangements at McIvers were defended by a series of tactics and strategies which responded to specific events and circumstances – the initial complaint, the State election, the local Council's budget for maintenance of public facilities, the technicalities of existing anti-discrimination legislation, etc. Nonetheless, the analysis of such struggles and practices, in all their messy empirical detail, can inform our reflections on struggles over the urban conditions of public-making.

My analysis of the struggle over McIvers suggests that 'exclusion' is a more complex problem for critical urban theory than it often seems to appear. Certainly, it suggests that exclusion is not a problem *per se*, and that critical urban theory must have the capacity to tell the difference between different instances of exclusion. In principle, some kinds of exclusion might be justified on the grounds that they facilitate the exploration of forms of co-presence and public sociability which are not possible in other public spaces. Given the spatiality of processes of public formation, the goal of a public sphere open to heterogeneity and diversity cannot be realized solely through the provision of spaces of interaction and mixing which are open to all. As Cooper (1998: 485) has argued:

> a heterogeneous approach [to space] does not mean – cannot mean – that all uses should be encouraged or permitted. Political decisions need to identify which uses, symbols, and activities should be restricted or excluded.

Rather than opposing all forms of exclusionary urban space, we might do better to focus critical inquiries both on the particular forms of publicness that are facilitated by instances of exclusion, and on the scope and accessibility of the debate through which it is justified and maintained. What distinguishes the exclusion sustained at McIvers from other forms of exclusion explored in previous chapters (such as The Northbridge Cufow) is that it is politically justified through a public debate about values rather than through the imposition of values that are assumed to be pre-political and universely shared by 'the public' and/or 'the community'.

Of course, exclusionary access might take all manner of forms. In the specific instance of McIvers Baths, a far-reaching and permanent form of exclusion has been maintained. An identity (men), rather than a use, symbol or activity, is excluded from the space. There are plenty of good reasons to be suspicious about attempts to establish identity through territorial boundary-making. For marginalized groups, the move from the tactic of 'temporarily occupying' to a strategy of 'territorially claiming' (Morris 1992: 28) brings with it all sorts of potential pitfalls. No attempt to create 'separate spaces' (be they 'women's spaces' or 'queer spaces', etc.) can ever be complete – like all attempts to carve a permanent identity out of the flow of relations which produce places, they will internalize their exterior to some extent. The sanctity of McIvers as a women's space is punctured by the occasional man who tries to enter the space. More than that, its very identity is partly constituted by an external environment in which certain forms of male gaze and violence are taken for granted. To that extent, this strategy risks reifying these external threats, and formal exclusions may thus 'lead the space to be seen as constituted by that which is forbidden' (Cooper 1998: 480). As Watson and Gibson (1995: 257) have noted:

'Walls' – or bounded spaces – occupied by specific groups may offer protection or places of resistance. These may be necessary for minorities to establish themselves. But even these spaces can quickly shift into places open to attack or abuse or lack of opportunity.

So, while Young (2000: 221) has argued in favour of the political value of 'differentiated solidarity' established through city spaces, she has also argued that any spatial and social clustering should not be 'based on acts of exclusion, but rather on affinity attraction' (2000: 224). For her, the zones in between such spaces ought to take the form of 'spatial shadings' and 'hybrid spaces' rather than fixed boundaries. Podmore's (2001) work on Montreal's Blvd. St-Laurent provides one example. She has identified the ways in which the 'shared space' of this street has facilitated the production of urban lesbian communities which are not territorial in the conventional sense. It is precisely because the street serves as a resource for many subcultural groups that lesbians feel able to blend into the crowd. As a result of this accessibility, the street:

facilitates patterns of sociability and communality, place-making strategies and even the expression of desire – despite the fact that it is not a 'lesbian territory' (Podmore 2001: 351).

I would certainly agree that the construction and defence of 'separate spaces' of withdrawal is not the only or the best way to establish and protect alternative forms of publicness. But there is no reason to think that attempts to manage a threat by externalizing it cannot co-exist with attempts to transform more open public spaces. Further, what interests me about the boundary which gives McIvers its identity is the fluidity which this fixity allows. In Michel de Certeau's (1984) terms, the permanent exclusion of men from McIvers has made it a 'proper place' for women, where certain expectations about behaviour and identity are sustained. However, the ongoing exclusion of men cannot be read as a complete or brutal 'triumph of place over time'. Even as men are permanently excluded, the forms of co-presence and sociability within McIvers change over time. Socio-spatial norms such as those sustained at McIvers are subject to tactical evasions and appropriations. And of course, these norms are also the object of contest and debate within the very counterpublic sphere through which they are established. So, while the exclusion of men is justified on the basis of some recognition of women's different needs or interests, this recognition need not fix or reify 'woman' as an identity category. Rather, it might facilitate an exploration of different ways of being together with (particular) strangers. Given the power differentials which characterize the production of public spaces, 'affinity attraction' alone might not be enough to secure spaces of difference.

Perhaps most importantly, in the case of McIvers the creation of a 'separate space' must be distinguished from the construction of the women's counterpublic as a 'separate camp' (Negt and Kluge 1993, see also Chapter 2). The campaigners successfully argued that in existing conditions of inequality, 'separate spaces' can serve as protected spaces where people can take the risk of exploring different ways of 'being together with strangers' with relative safety from attack and abuse. But their argument for the exclusion of men was justified through a political and legal campaign which engaged with other publics, and made its claims with reference to concepts of justice and equality. Here, we see a counterpublic sphere which is produced through a separate space but not as a separate camp.[5] As Phil Hubbard (2002: 66) suggests, we may sometimes do better to think of 'zones of alternative citizenship' as 'critically exclusive' rather than 'radically inclusive'.

Chapter Eight

Imagining the Public City: Concluding Reflections

One night in 1994, I sat down at home to watch a national current affairs programme called *The Times*, which was due to televise a story about the first national conference of police and young people from non-English-speaking backgrounds. At the time, I was working for the Youth Action and Policy Association of NSW (YAPA), a peak body for community-based youth services and young people's organizations. YAPA was one of a number of organizations that had expressed concern about over-zealous and racist policing of young people from non-English-speaking backgrounds in Sydney (Youth Action and Policy Association of NSW and Youth Justice Coalition 1994). The purpose of the conference, organized by police, was to address such criticism and to find ways forward in collaboration with young people from around the country.

The Times had been invited to cover the conference by police, who were keen to display their willingness to listen to young people. The story began with happy pictures from the conference. 'On the surface,' *The Times* observed, 'things were warm and fuzzy'. The Assistant Police Commissioner from Western Australia was 'quite frankly amazed at the calibre and quality of the young people who came forward'. One of the young delegates hoped that the conference would be the first step in a 'healing process' between young people and police. The gathering ended with music and hugs (really).

The scene then changed to a pedestrian mall in Bankstown, a suburb in the southwest of Sydney. There, *The Times* interviewed the group of young people from New South Wales who had attended the conference, to ask them about the issues that had been raised and their hopes for the future. I had worked closely with these young people and knew some of them pretty well, so this was the bit of the story I was excited to see. One of the interviewees quipped that, as a group of ethnic minority young people on the street, they were bound to attract the attention of police. Sure enough, viewers were told that within the first thirty minutes of their interview, four police cars had rolled through the mall to observe proceedings. Eventually, a police wagon stopped. An officer approached two young men who were friends of the group being interviewed and were watching the interview at a distance. The cameras swung around to focus on the action as the police officer questioned one of the young men – a young man of Lebanese descent. The interviewees explained to *The Times* that this was exactly the kind of police treatment that many young people had come to expect. The young man was asked to raise his arms and spread his legs, and he was searched. He was then told to empty the contents of his pockets for the police officer. The police officer found $100 cash. He then placed the young man under arrest.

As the young man was escorted into the back of the police wagon, his friends protested to police. Why was he being arrested? What was he supposed to have done wrong? Their questions were ignored, the atmosphere highly charged with confusion and anger. The young man's girlfriend, who had attended the conference, asked through the closed door of the wagon if he knew why he had been arrested. He replied, 'fuck, I dunno, it's something about the money in my wallet'. For this comment, he was later charged with offensive language in a public place. No other charges were laid.

Over a decade has passed since this event, but it sticks in my mind as a moment when a penny dropped for me – I can pretty much date my interest in urban public space to this night. After I'd stopped shouting at the television, I started to wonder whether this kind of incident might signal something bigger than the existence of a problem between young people and police – perhaps it signalled something about the changing nature of urban life more generally? When I began researching this matter, I soon discovered a number of other critical urban theorists had already come to this conclusion. My awakening to these issues happened some four years after the publication of Mike Davis's (1990) influential book *City of Quartz*, and two years after the publication of Michael Sorkin's (1992) edited collection *Variations on a Theme Park: The New American City and the End of Public Space*. Some Australian scholars had already started to ask whether the fortification and privatization of public space witnessed in American cities was under way locally (see for example Davison 1994), and the exclusionary policing of Bankstown mall certainly seemed to support such a thesis.

Such analyses of public space suggested that the kind of harassment suffered by this young man was becoming an everyday occurrence for many people in cities in Australia and beyond. Those whose identities are despised or feared, who are constructed as 'different' or 'other', seemed to have their access to public space restricted by zealous policing (by both state police and private security guards) supported by surveillance, exclusionary planning regulations and restrictive property regimes. Even if such groups had achieved formal citizenship rights through membership of a nation-state, this seemed no guarantee of substantive citizenship rights – that 'array of civil, political, socio-economic, and cultural rights' which are so constitutive of everyday life (Holston and Appadurai 1996: 190). As Joe Painter and Chris Philo (1995: 115) put it:

> If people cannot be present in public spaces (streets, squares, parks, cinemas, churches, town halls) without feeling uncomfortable, victimized and basically 'out of place', then it must be questionable whether or not these people can be regarded as citizens at all: or, at least, whether they will regard themselves as full citizens on an equal footing with other people who seem perfectly 'at home' when moving about in public spaces.

Contesting the treatment of my friend in Bankstown mall, then, seemed to me at first to be a matter of defending 'public space' against exclusionary processes of fortification and privatization.

As we saw in Chapter 1, the concept of 'public space' is frequently deployed as a spatial imaginary in order to identify and oppose instances of urban injustice such as the incident screened by *The Times*. Visions of urban public spaces which are 'open to all', 'real' and/or 'democratic' are contrasted with actually existing urban spaces that are 'closed', 'fake' and 'authoritarian'.

I am no longer convinced as to the political utility of this spatial imaginary. If one of the purposes of critical urban theory is to identify and oppose those actions which unjustly constrain the making of publics, I do not believe that this project is best served by conventional conceptions of public space. To frame our analysis of public urban geographies with reference to topographical conceptions of public space is to blinker our vision – such analyses start by paying attention only to particular places (those places that are or ought to be 'public space'), and then asking about their accessibility – are they 'open to all'? In this book, by contrast, I have argued that the 'how' and 'where' of public address are dynamic rather than fixed. Precisely because publicness takes many forms and has many geographies, there is no exclusive or privileged connection between public address and 'public space' (in its conventional topographical sense). 'The public', as Clive Barnett (forthcoming) argues, 'is not to be found anywhere special, it has no proper place, nor any exemplary spatiality'. In this context, the challenge for critical urban theory is to develop a political and spatial imaginary which is alive to the multiple dimensions of both publicness and the city. In Chapter 2, I set about articulating an alternative understanding of the urban dimensions of public-making. In subsequent chapters, I have applied the three-pronged framework developed there to the study of five struggles over the conditions in which people have attempted to come together as a public. In concluding this book, I want to argue that the framework I have developed and applied across this book suggests that our imaginations of the public city must recognize that: (a) the city is not (just) a 'stage' for theatrical forms of public address, but is mobilized for different forms of public address which often involve combinations of co-present and mediated interaction; (b) the visibility associated with being 'in public' is simultaneously a resource and an impediment for public-making, and different forms of publicness mobilize different combinations of visibility and invisibility in response to this tension; (c) the 'public city' is a product of the particular political labours which seek to make particular publics, rather than the product of a shared commitment to a normative ideal of city life as the 'being-together of strangers'.

The City Is Not (Just) a Stage

'Modern architecture only becomes modern with its engagement with the media', argues architectural theorist Beatriz Colomina (1996: 14). The same can be said, I think, of the modern city. We simply cannot grasp the urban dimensions of public address through frameworks which assert that the city's 'public spaces' are preferable alternatives to mediatized public spaces, nor through frameworks which assert that the city has been rendered irrelevant by the media. To the extent that urban sites are productive of different forms of publicness, this occurs through their relationship to other media of public address. Public address mobilizes different spaces and spatialities in a variety of combinations which require careful analysis.

For a long time, I remained blind to the importance of such combinations in my thinking about what happened in Bankstown mall – I simply did not 'see' the television cameras. But it now seems to me that the presence of the cameras, and the screening of this incident, are absolutely crucial to understanding what happened, for at least two reasons. First, the presence of the cameras undoubtedly shaped the actions of those involved. Most obviously, the young people were in the mall in the first place in order to appear before the cameras. And it is at least plausible to assume that the police officers involved might have arrested the young man as a display for the television crew. Nightly current affairs programmes in Sydney and other Australian cities regularly screen stories about 'streets out of control', complaining about lack of police action against the 'anti-social' who loiter and threaten 'the community'. Making an arrest in front of the cameras would surely make for some positive media coverage for once (wrong!). Second, the presence of the cameras and the subsequent screening of the incident were fundamental to the 'publicness' of this incident – in the sense that they provided the young man a further opportunity to contest his treatment in Bankstown mall by addressing a wider audience. This was not inconsequential to the final outcome. The story was screened nationally on the night that the incident occurred. Sydney lawyer and newspaper columnist Chris Murphy happened to be among the 'strangers' watching *The Times* that night. He represented the young man for free in court, and the charge was dismissed. *The Times* followed the case, and the norms of occupation of Bankstown mall became a matter of interest to a public beyond those who were physically present to 'witness' the incident as it took place.

What I missed in my initial consideration of these events, then, was 'the story' itself, and the different spaces through which the young man circulated claims about his treatment. His arrest was visible to those present in the mall and those who saw it 'on screen'. And of course, in subsequently relating this story myself in conferences and seminars and now in this book, I am further contributing to the public circulation and discussion of this incident. To the extent that what transpired in Bankstown was a public event, then, this was certainly not simply a matter of it occurring 'in public', where 'in public' is understood in a theatrical sense of an audience gathered together at a particular time to witness a performance in a shared physical space. The publicness of the event was (and is) actively made through actions across a range of spaces and media. Colomina (1996) is right to argue that the audience of a public event should not be confused with the audience that gathers together in an urban 'public space'. To be seen, for Colomina (1996: 7–8):

> no longer has much to do with a public space, in the traditional sense of a
> public forum, a square, or the crowd that gathers around a speaker in such
> a place, but with the audience that each medium of publication reaches, independent of the place this audience might actually be occupying.[1]

In missing 'the story', then, I missed a vitally important aspect of this incident. While the young man and his advocates acted across various spaces and media to publicize the incident and contest his treatment, in my mind's eye he was fixed in the back of a police wagon.

Of course, the rapid diffusion of new media and communications technologies in the latter half of the twentieth century has initiated plenty of discussion concerning the ongoing relevance of the city to public life. Anxious to defend the city against charges of its irrelevance, some have claimed that the forms of public address sustained in and through the city are *preferable alternatives* to the forms of public address sustained in and through the media. This defence of the city's importance to public life is frequently made on the grounds that embodied co-presence in 'public space' is the best possible condition for public-making. Here, publics are imagined in theatrical terms: the public spaces of the city are envisaged as a stage from which speakers/performers can address an audience assembled before them. Such 'unmediated' encounters are said to sustain the best possibility for creating strong and active publics – in a shared 'public space', people have the opportunity to put their case to an audience directly, and the audience has the potential to respond collectively and immediately. Richard Butsch (2000: 15) puts it this way:

> Critical to any conception of public sphere and also to any potential for collective action is conversation, for the opportunity to assemble and discuss and come to consensus about what to do.

Advocates of co-present forms of public address like Butsch are critical of the mediated interactions of print, radio, television and the internet, which are said to privilege passive forms of public life where strangers are isolated from one another. This isolation is said to make collective public action more difficult, and to allow people to avoid those who are different by simply putting down the newspaper, turning a dial, flicking a remote or clicking a mouse. 'Screening' has a double meaning – to screen can be to 'screen out', and this is suggestive of the inequality of access that permeates media spaces. So, if the screen has become the dominant space of public address, then this is to the detriment of public life.

In my view, the claim that 'urban' co-present forms of public address are preferable alternatives to mediatized forms of public address is unsustainable. There are two main problems. First, while I think critics of mediatized and screened forms of publicness are right to point out their limits, theatrical and staged forms of publicness also have their limitations. Like 'screening', 'staging' also has a double meaning – the staging of public events may be calculated to manipulate audiences, 'stage-managed' as it were. In Perth's Central Area, stages are regularly erected in order to parade fashions and other products on sale at retail outlets. At Parliament House, state functions are regularly staged before an assembled, hand-picked audience. Further, the vision of public-making as a process of theatrical assembly is simply not feasible as a model for contemporary democracies. As Young (2000: 46) points out, this imaginary of 'centred' public interaction 'implicitly thinks of the democratic process as one big meeting at the conclusion of which decisions are made, we hope justly'. Certainly, she argues, face-to-face discussion is important. However:

> A discussion-based democratic theory will be irrelevant to contemporary society . . . unless it can apply its values, norms, and insights to large-scale

politics of millions of people linked by dense social and economic processes and legal framework . . . The challenge for a theory of discussion-based democracy is to explain how its norms and values can apply to mass polities where the relations among members are complexly mediated rather than direct and face-to-face (Young 2000: 45).

The construction of a public sphere on this scale relies upon the knitting together of public action across a range of spaces into a 'larger space of nonassembly' (Taylor 2004: 86). In this context, it is hard to see how face-to-face interactions in urban public spaces are *preferable* to mediatized interactions – both have their limitations.

Second, the 'stage' and the 'screen' (or, for that matter, the 'polis' and 'print') should not be seen as mutually exclusive *alternatives*, as if they were enemies locked in a battle for historical and geographical ascendancy. New forms of public address do not replace the old, nor do the old forms survive by avoiding contamination by the new forms in order to remain pure. Such oppositions simply do not hold when we explore the dynamics of different forms of public address and their connection to the urban. My investigations into the geographies of public address demonstrate that those engaged in the making of publics often combine actions on stage, in print and on screen in crafting scenes of circulation. Protests at Parliament House in Canberra are in many instances organized and structured in order to access direct witnesses *and* to achieve newspaper and television coverage – indeed, sometimes their main purpose is to achieve media coverage. Further, some protests incorporate media technologies into the event itself, using screens and public address systems to magnify and amplify speakers.[2] Graffiti writers often select particular urban surfaces for their capacity to be enable the further circulation of graffiti texts through coverage in newspapers, magazines, television shows, movies and websites. Even the fleshy encounters at beats are increasingly informed by mediatized discussions of beat etiquette and safety in the gay and lesbian press and encounters in internet chat rooms. In each of these instances, public action involving co-present interaction in particular urban sites is shaped by, and/or geared towards, public action in mediatized spaces. Conversely, in each of these instances the mediatized spaces of public address such as television and cyberspace are dependent upon (and therefore shaped by) action staged in sites in the city. City and media take shape in relation to each other.

In arguing that co-present forms of public address are neither preferable nor mutually exclusive alternatives to mediatized forms of public address, I do not mean to suggest that all forms and spaces of public address are either identical or equivalent. Different kinds of 'public space' (procedurally defined) offer different kinds of possibilities for public address.

These differences matter because, as Clive Barnett (forthcoming) puts it, public action is 'parasitical'. That is, public address takes shape through:

> the material configurations and social relations laid down by other forms of activity, in the sense that it is dependent on these as its conditions of possibility, as well as in the sense that it is these conditions that in turn become the object of transformative public action.

In this context, we must ask: 'how does the public dimension of discourse come about differently in different contexts of mediation?' (Warner 2002: 162). This question is frequently a topic of debate for different publics, as participants experiment with various combinations of spaces and forms in order to circulate texts. For example, some protestors expressed concern that the regulation of protest at Parliament House has the effect of privileging mediated public address over direct public address. Some graffiti writers expressed concern over the proliferation of information about graffiti made available as graffiti-related magazines and internet websites grow in popularity. But from my perspective, such debates about the relative merits of different spaces and forms of public address are not best settled with reference to ontological claims which insist that their different 'shapes' *determine* the quality and styles of public discourse they can sustain. Rather, we need to investigate carefully the circumstances in which choices about spaces and modes of public address are made (and constrained) by participants in different public spheres.

My argument here concerning the relationship between stage, print and screen is similar to (and influenced by) Philip Auslander's (1999) analysis of 'liveness' as a modality of performance. He notes that in contemporary performance studies, the 'liveness' of theatre is often held in higher esteem than mediatized performance by those who insist that the theatre has more radical political potential than mass media. But the very concept of 'liveness', says Auslander, is a function of mediatization. This is true in more than one sense. First, 'live performance now often incorporates mediatization such that the live event itself is a product of media technologies' (1999: 24). Second, live events are often 'pre-adapted (so to say) to the demands of their new medium' (1999: 27). Thus, for him, 'mediatization is now explicitly and implicitly embedded within the live experience' (1999: 31), in both technological and epistemological senses. The implication of this, says Auslander (1999: 51), is that:

> To understand the relationship between live and mediatized forms, it is necessary to investigate the relationship as historical and contingent, not as ontologically given or technologically determined.

It is not that there are not important differences between live and media-tized forms. Rather:

> any distinction needs to derive from careful consideration of how the rela-tionship between the live and the mediatized is articulated in particular cases, not from a set of assumptions that constructs the relation between live and mediatized representations *a priori* as a relation of essential opposition (1999: 54).

Likewise, my study of the urban dimensions of struggles over public address has sought to investigate how different spaces and modes of public address are configured in relation to one another in the efforts of particular people to make publics, and how the urban penetrates (and is penetrated by) each of these spaces and modes. Our understanding of the urban dimensions of public address is held back by the persistence of visions of the city as a stage for unmediated forms of public address which operates as a preferable alternative to media spaces.

If we open up our understanding of the different combinations of spaces that are put to work for public address, then, we become more sensitive to the possibilities that these various spaces afford. But we must also ask, possibilities for what? Whether we think of public urban geographies in terms of 'stages' and 'screens', our ways of talking about publicness often suggest that being public is a matter of *appearing* before others. I now turn my attention to the politics of public appearance and its connection to the urban.

Seen and Unseen

> We are much less Greeks than we believe. We are neither in the amphitheatre, nor on the stage, but in the panoptic machine, invested by its effects of power, which we bring to ourselves since we are part of its mechanism.
> (Foucault 1977 *Discipline and Punish*: 217)

My second set of conclusions concerns the relation between the two dimensions of publicness – visibility and collectivity, or *being in public* and *being public* – and their connection to urban life. My contention here is that we must be alive to the *tensions* between being *in* public and being public. Appearance in public facilitates both the circulation of texts *and* the normalization of behaviour, both subjecthood *and* subjectification. This has significant implications for our discussion of the urban dimensions of struggles over the making of publics. Put simply, if being 'in public' offers possibilities for being seen, then this visibility is potentially both a resource for, *and an impediment to*, public-making. Managing this tension is a necessity for the making of counterpublic scenes of circulation, as we have seen in the studies of public-making presented in the preceding chapters. 'Appearance' can facilitate the circulation of texts *and* social control. In this context, neither being seen nor being unseen is to be celebrated without reservation.

In the opening chapters of this book, I defined 'being public' as a matter of participating in a 'scene of circulation', through which one could circulate a text to strangers. In this understanding of publics as scenes of circulation, participation in a public is premised on one's capacity to establish oneself *as a participant* in that scene – that is to say, participation in a public is not solely a matter of being, but a matter of being public. Visual metaphors are frequently deployed to conceptualize this process of establishing oneself as a participant in a public scene. Hannah Arendt (1958: 38, 51), for example, equates being public with *appearance* before others in a public realm distinguished from the private realm by its 'merciless exposure' and its 'harsher light'. As such, it is not surprising that questions about access to these scenes of circulation are framed in terms of visibility: one who is restricted to the invisibility of privacy suffers a form of deprivation:

> To live an entirely private life means above all to be deprived of things essential to a truly human life: to be deprived of the reality that comes from being seen and heard by others . . . The privation of privacy lies in the absence of others; as far as they are concerned, private man does not appear, and therefore it is as though he did not exist (Arendt 1958: 58).

From this perspective, being part of a public (in the collective sense) would certainly seem to be a matter of being in public (in the visible sense). Publicness as visibility reinforces, even acts as a condition of, publicness as collectivity. This is why the city's contribution to the making of public spheres is frequently associated with the opportunities for being seen (see for example Mitchell 2003).[3] Here, 'invisibility' is cast as a problem to be overcome – Sandercock (1998), for example, talks of the need for radical urban planning practice to make 'the hitherto invisible visible' in order to 'make it discussable' in wider public debate.

However, the visibility associated with being 'in public' has another set of effects beyond enabling participation. Most importantly, fields of visibility are bound up with the techniques and technologies of discipline through which the some ways of being (in) public are normalized. To be 'in public' is to have one's conduct exposed to the normative gaze of others, and exposure to this gaze is one of the technologies of governance which incite us to regulate our own conduct with regard to what is 'appropriate' when in public. As Foucault put it in his discussion of 'The Eye of Power' (1980: 154), through the reign of 'opinion':

power will be exercised by virtue of the mere fact of things being known and people seen in a sort of immediate, collective and anonymous gaze. A form of power whose main instance is that of opinion will refuse to tolerate areas of darkness.[4]

It is because of this aspect of being 'in public' that Arendt (1958) speaks of the light of publicity as 'harsh' and 'merciless'. Indeed, Arendt's discussion of the rise of 'the social' as a distinct mode of publicness was intended to capture precisely this normalizing dimension of public visibility.[5] In her analysis, modern 'society':

> expects from each of its members a certain kind of behaviour, imposing innumerable and various rules, all of which tend to 'normalize' its members, to make them behave, to exclude spontaneous action or outstanding achievement (Arendt 1958: 40).

The young man arrested in Bankstown mall would have been only too aware of this disciplining aspect of public visibility. His age and the colour of his skin and his style of dress made him highly visible to police, who are prone to 'profiling' the likely culprits of 'anti-social behaviour' which has been defined in opposition to normative forms of urban sociability. Here, his appearance fit the established stereotype of someone who is outside of 'the social', literally 'anti-social'.

One of the interesting points of comparison across the case studies presented in preceding chapters lies in the different procedures adopted by participants in counterpublics for managing the tension between the circulatory and disciplinary effects of visibility. Indeed, the styles of public address sustained by these counterpublics are defined by their particular combinations of visibility *and* invisibility. We can detect at least four different procedures for combining the visibility and invisibility required for the making of counterpublics. First, as counterpublics combine different kinds of spaces, these spaces are invested with different roles – some spaces and times are mobilized for performances exposed to a wide audience, while other spaces and times are mobilized for preparation of such performances. We might use Goffman's (1959) famous distinction to describe such spaces as 'frontstage' and 'backstage'. Or, in light of my earlier concerns about 'stage' metaphors for urban public address, we could equally liken this to the distinction between performance on screen and off screen, or between writing a published work and personal diary. We can think here of protestors planning their actions at Parliament House in small gatherings at offices, meeting halls, lounge rooms and internet chat rooms. Or we might think of graffiti writers working on outlines in their sketch books before bringing them to life on a wall.

Second, while the language of 'stage' and 'backstage' suggests that visibility and invisibility are associated with different spaces, counterpublics have also crafted combinations of visibility and invisibility by making use of a given space at particular times. The night, for example, has provided a cloak of reduced visibility for experiments with different forms of public address. Graffiti writers have historically made use of the darkness of night to write their pieces, perhaps hoping that the 'eye of power' will be drowsy at certain times. Of course, the finished product is quite visible during the daylight hours, at least until it is removed. Indeed, graffiti could be described as a form of 'night discourse', where the night-time city speaks back to the day-time city (Cresswell 1998). Much beat activity (though by no means all) also takes place at night. Given the ways in which the night-time city is mobilized for the making of counterpublic scenes of circulation, it is of no surprise that the darkness of night is often conceived of as a problem by urban authorities. Indeed, the history of street-lighting is bound up with the history of urban policing (Schivelbusch 1987), and lights continue to be mobilized in efforts to prevent graffiti-writing and beat sex (among other forms of behaviour). And in Perth, of course, we have seen the imposition of night-time curfews in efforts to create and protect a normative model of 'public space'.

Third, participants in some counterpublics have become experts at 'blending in', combining visibility and invisibility simultaneously in the same space and time. To use Dick Hebdige's (1988) wonderfully apt term, some urban practices 'hide in the light' of publicity, conforming to normative expectations in appearance only to avoid them in some way. Consider, for example, the actions of the choir of women at Parliament House in Canberra and cruising men. In both instances, their capacity to 'blend in' was/is fundamental to the opportunities they craft for particular styles of public address – the women blended in as visitors to Parliament House, cruising men blend in as men simply walking Blanche the dog in a park, and yet each found ways to address others through non-normative forms of publicness. Similarly, some graffiti writers in Sydney and elsewhere have begun to write graffiti and alter commercial billboards dressed in overalls and hard hats in the middle of the day, finding this to be an effective way of avoiding suspicion – they blend in as workers attending to the business of replacing one urban image with another (see for example *Sydney Morning Herald*, 30 September 2005, p. 3).[6] As Colomina (1996: 11) has put it, 'Sometimes the best way to hide something is in full sight'.

Fourth, participants in one counterpublic under examination here made use of visible physical barriers to establish productive forms of invisibility. At McIver's Baths, fences and rules are deployed on behalf of women users who do not want to be seen by men. In this instance, while the layout and the rules of access to the Baths prevented users' exposure to the direct gaze of men, both the layout and the rules became highly visible in mediatized discussions associated with the campaign to protect McIvers from claims of unlawful discrimination. Of course, the boundaries established by fences and rules are themselves subject to leakage – some men have had to be chased out of McIvers during the day, while others take illicit dips there during the night.

In their different ways, these different counterpublic procedures show that society has not been entirely reduced to the 'grid of discipline' (Certeau 1984: xiv), and power is not always victorious (Foucault 1980: 163–4). The city, then, is more than a 'panoptic machine'. The 'eye of power' is neither all-seeing nor all-powerful, and areas of relative 'darkness' persist. Of course, it is precisely because of the disciplining effects of being seen 'in public' that writers like Michel de Certeau (1984) have celebrated the invisibility of those everyday urban practices which evade domination by remaining hidden from (or at least opaque to) the 'eye(s) of power'. He celebrates those 'procedures that elude discipline without being outside the field in which it is exercised' which are able to prise open a 'crack in the system that saturates places with signification' (Certeau 1984: 96, 106).

But we should not simply celebrate 'invisibility' as a form of resistance to domination through the norm. Invisibility might be externally enforced, it might be established through tactical evasions and appropriations, and it might be achieved through strategic enforcements. It is also not equally accessible to all, in situations where visible bodily attributes serve as markers for harmful stereotypes. Furthermore, it would be a mistake to associate counterpublics exclusively with 'tactical' forms of invisibility and dominant publics with 'strategic' forms of visibility.[7] As Meaghan Morris (1992) has observed:

> The notion of tactics is not as romantic as it can sound. For one thing, it is parasitic on the notion of 'strategy': if this means that 'tactics' cannot of itself sustain a theory of autonomy for strategically 'othered' people, it also means that it cannot be used to ground an ontology of otherness (and that individuals cannot be treated holistically as full subjects of either strategy or tactics).

No public sphere, including counterpublic spheres, could be constructed by people who remain in a state of total invisibility. So, there would not be much hope for the formation of counterpublics if total invisibility was a condition of their existence. In any event, such invisibility is increasingly difficult to achieve. As Thrift (2004: 53) notes, new ways of gathering and processing 'locationally referenced information about everyday life' mean that 'most of the spaces of everyday life will no longer be hidden at all'. In these circumstances, 'it is not so much hiding as trying to fashion different modes of visibility that is crucial' (Thrift 2004: 53–4).

The project of fashioning these 'modes of visibility' is a fraught process. In particular, when public claims are made in an effort to secure a 'proper place' for counterpublic practices, this inevitably renders them more visible. The resultant visibility of such practices has the potential to undermine the very conditions which sustained them in the first place. Claims on behalf of legal graffiti contributed to the wider circulation of knowledge about hip hop culture, and to the subsequent deployment of this knowledge by police and other urban authorities attempting to close down opportunities for writing graffiti. The Beat Project's promotion of safe beat sex made the use of beats more visible to council workers and police. A successful claim to sustain a proper place for women at McIvers raised the visibility of the pool, and led to an increase in men perving on the place and its users from the cliffs above. Nonetheless, if people want to contest the status of their counterpublic sphere by taking political action, they cannot avoid some degree of visibility, regardless of the potential costs and risks that this involves. While de Certeau (1984) might have celebrated the 'ordinary man' of everyday tactical practices on the grounds that he 'comes before texts' and 'does not expect representations', political struggles to publicly contest 'the norm' cannot exist in a state of invisibility. As Susan Ruddick (1996: 195) has concluded, a failure to move beyond celebrations of tactical invisibility:

> presents us with a danger far greater than that of simply describing how power is exerted over marginalized peoples. To speak exclusively of tactical forms of resistance is to risk normalizing, even romanticizing, the condition of the marginalized people, humanizing the face of poverty in a way that demands no further action.

In this context, our imagination of the public city must be finely attuned to the different 'modes of visibility' that are formative of different ways of being public, rather than directly equating being public with being seen (or unseen) *in public*.

Strangers on the Horizon

My final set of reflections concerns the construction of 'the city' as a social imaginary. What does our analysis of the urban dimensions of public-making tell us about 'the city' as a context for address and interactions among 'strangers'? How should we imagine the 'public city'?

In *Justice and the Politics of Difference*, Iris Marion Young (1990) famously proposed that our experience of city life is suggestive of a normative ideal for the 'being-together of strangers' in a heterogeneous public sphere. 'City dwelling,' she argued, 'situates one's own activity in relation to a horizon of a vast variety of other activity, and the awareness that this unknown, unfamiliar activity affects the conditions of one's own' (1990: 238). The element of city life that inspires Young is the capacity of city dwellers to recognize that *their ways* of being together as a public are not the *only ways* of being together as a public. For Young, this recognition opens up the possibility that different publics might somehow find a way to coexist in a polity characterized by differentiation without exclusion. Young's influential formulation continues to be a useful departure point for further reflections on the urban imaginary, because it directs our attention to the centrality of relations among strangers for our understanding of the public imaginary and the urban imaginary (see also Donald 1999; Harvey 1997; Sandercock 1998, 2003; Young 2000).

Of course, Young realized that while a normative ideal for the interaction of multiple publics might have been suggested by our experience of urban life, this ideal remains largely unrealized. As we have seen in previous chapters, 'the city' is all too often imagined by participants in some publics to be a social totality which is simply an extension of their particular public. As such, they imagine their own particular norms of public address and sociability to be *the* norms of public address and sociability. They consider people who are unable or unwilling to embrace their values concerning the kinds of behaviour that are proper in 'the city' to be 'anti-social'. Difference, from this perspective, becomes a matter for the 'police'. The 'police', here, refers not (only) to those police who wear uniforms, such as the officers who removed the young man from Bankstown mall and the 'JAG Team' which enforces the curfew in Perth. Rather, I am using 'police' to refer to a wider set of procedures which seek to allocate particular ways of being and particular bodies to their 'proper places' on behalf of 'the city' (Rancière 1999: 28–9). So, when protestors refuse to stick to the rules of protest, when men cruise in parks, when young people write graffiti and hang out together in noisy groups, they are policed – by which I mean that they are subject to a variety of efforts to assign them a proper place and to keep them in their place. The police might be wearing uniforms and defending 'the community' against the 'anti-social', or they might be expert planners making technical determinations about land use functions on behalf of 'the public interest' they claim to represent.

In this context, the question for those concerned with the relationship between publics and the city is this: how might the *policing* of difference be made the object of a *politics* of difference, so that the possibilities for sustaining different forms of publicness in and through the city are politically *debated* by urban inhabitants rather than *determined* by urban authorities? Visions of the public city as a 'being-together of strangers' are appealing precisely because they promise to give us both critical and political purchase on this question. Critically, this vision of the public city offers a theoretical tool for *diagnosing* the ills of actually-existing urban life across a wide range of contexts. So, for instance, this vision of the public city could be mobilized as the basis for a critique of the arrest of my friend in Bankstown mall. From this perspective, his arrest could be viewed as but one example of the hostility towards 'strangers' which prevents city life from living up to its promise as a 'being-together of strangers'. To the extent that the police who made the arrest claimed to be protecting a 'community' whose values are assumed to be beyond dispute, their actions were problematic – the vision of the public city as a 'being-together of strangers' insists that there can be no set of values or standards for living together which are beyond political contestation. Politically, this vision of the public city offers a *cure* for the ills of urban life which also promises to have applicability across a wide range of contexts. Because city life is by definition a 'being-together of strangers', living in the city is said to demand that urban inhabitants be 'reasonable' in their dealings with one another. That is, it is argued that a genuine politics of difference can emerge only if urbanites imagine themselves to be living together in the city *as strangers*. The realization of the ideal public city is premised on the capacity of urban inhabitants to recognize their own particularity and consider their own needs and interests in relation to the needs and interests of others. As we saw in Chapter 2, this demand for urban inhabitants to be 'reasonable' or 'open' to one another has been advanced by Young (2000) and a range of other urban theorists (Deutsche 1999; Donald 1999; Sandercock 2003; Tajbakhsh 2001).[8]

The appeal and the strength of this vision of the public city lie in the productive assertion that there can be no set of standards for living together in the city outside of politics. But this vision of the public city also has some significant limitations. One immediate problem with demands for urban inhabitants to be 'reasonable' by treating themselves and others as strangers is that such demands are likely to be ignored by the dominant publics to whom they are implicitly addressed. As Susan Fainstein (1999: 256) has observed, demands for urbanites to adopt a 'reasonable' or 'open' stance towards strangers are likely to fall on deaf ears:

> In the face of deepening economic polarization of restructured capitalism, the call for respectful discussion is unlikely to provoke a meaningful response from the possessors of social power.

Of course, none of the theorists mentioned above would be so idealistic as to believe that city life will be changed without political struggle. But a more significant problem with the vision of the public city is that it obscures (or even denigrates) some of the most important concerns of those who wage these struggles. These struggles, and public action in general, have a purposive dimension – they are about something (Barnett forthcoming). The political labours of the people discussed in this book were conducted on behalf of protest, and cruising, and graffiti, and hanging out with friends, and women swimming in the company of other women, and so on. But visions of the 'public city' as a 'being-together of strangers' have little to say about the instrumental, purposive dimensions of public action. Advocates of this vision of the public city tend to be more concerned with the intrinsic dimensions of public action – based on the claim that city life is by definition a 'being-together of strangers', they offer a normative ideal which is geared towards providing generalizable diagnoses and cures for urban life. In doing so, they tend to privilege a 'purist conception of 'the political' defined in stark opposition to more instrumental, purposive understandings of politics' (Barnett 2004: 191). Indeed, the concept of the public city as a 'being-together of strangers' and associated demands for urban inhabitants to be 'reasonable' stand outside of the very struggles that are conducted to sustain different publics in and through the city in pursuit of particular political goals.

What kind of vision of the public city might emerge if we focus on the instrumental, purposive nature of public action? A revised vision of the public city must have at its heart a concern with the political labours of people who engage in struggles to make particular publics in particular contexts. Such a revised vision of the public city will still be concerned with the nature of encounters between 'strangers'. But rather than deploying a vision of the city as a 'being-together of strangers' in order to articulate the terms on which such encounters ought to be conducted, we should situate these encounters in relation to the pragmatic, instrumental calculations which are made by those who struggle to make publics. In other words, we should heed Barnett's warning not to 'put the normative cart before the pragmatic horse' (Barnett 2004: 191). A closer engagement with the political labour of public-making suggests two important revisions to our understanding of encounters between 'strangers' in the public city.

First, a focus on the political labour of people who struggle to make publics in and through the urban does not support a vision of the public city as a 'being-together of strangers', in the sense that strangerhood is a universal condition shared by all. In fact, the making of publics in and through the urban involves efforts to *differentiate* between the strangers on one's horizon. Given that public-making proceeds in contexts where some strangers might indeed want to 'destroy and exterminate' (Kearney 2003: 10), it is not unreasonable for urban inhabitants to differentiate between the strangers on their horizon, and to make pragmatic calculations of how they might address some strangers while ignoring, avoiding or excluding others. To take but one example from the previous chapters, it would be impossible to sustain a cruising counterpublic sphere unless participants were able to differentiate between the strangers who are open to being cruised, the strangers who are indifferent to or ignorant of cruising, and the strangers who seek to shut down opportunities for cruising. Visions of the public city which demand that urbanites be 'reasonable' and 'open' to 'strangers' in general as a matter of normative first principle tell us little about how such pragmatic calculations are made. It seems naïve at best to suggest that urban inhabitants should act on the basis of a vision of the city as a 'being-together of strangers', when their experience of the strangers on their horizon is that not all will respond in the same way to finding themselves in the sphere of different styles of address. So, while they might expect some 'strangers' to want to participate in a particular scene, they might expect other 'strangers' to be indifferent to that scene, and they might imagine yet other 'strangers' to be hostile to the very existence of that scene. Calculations about the different strangers on the horizon of a public sphere are in fact constitutive of the styles of address that they sustain. Of course, these calculations may be proved wrong, and so every encounter with another has transformative potential (Ahmed 2000). But this does not imply that pragmatic attempts to differentiate the strangers on one's horizon are therefore inherently unjust – this implication could only be supported if the only thing at stake in struggles to make publics was the realization of a normative ideal of city life.

This brings me to my second observation about the public city, which concerns the circumstances in which people might come to imagine themselves as part of a 'city', which is understood as a shared horizon of address and object of public action. Rather than insisting that city life is intrinsically a 'being-together of strangers', we can also situate the production of specifically *urban* imaginaries in relation to the instrumental dimensions of publicness. I remain suspicious of any suggestion that people's efforts to frame their own particular needs and interests in relation to the needs and interests of 'the city' are a matter of ethical necessity. We might reconceptualize such efforts as the product of debate and reflection within particular counterpublics which are driven by pragmatic calculations about how they can best pursue their particular interests. That is, the question of whether or not one ought to imagine oneself to be part of 'the city' is itself a question for politics and public debate, not something that is settled in advance of politics. As we have seen, participants in counterpublics are constantly engaged in debates about how best to sustain their scenes of address and sociability. Some participants may attempt to transform the context in which they come together as a public by engaging with the wider horizon of 'the city'. Others have sought to insulate their public scenes from the indifference and hostility which they face. So, some frustrated protestors at Parliament House have played by the rules of protest while arguing before parliamentary committees for a rethink of the kinds of protest that are 'proper' to Canberra as the 'National Capital'. Others have thrown paper aeroplanes at politicians from the public galleries. Some graffiti writers have argued with politicians and forged alliances with sympathetic youth workers and academics in order to advocate the provision of legal opportunities for different forms of graffiti which make a proper place for graffiti in the city. Others develop their own ethics about where they ought to write which refuse to conform to capitalist property relations, and are of the opinion that mainstream society can go and get fucked. Some cruising men have volunteered their time and put themselves at considerable risk in providing health education and fighting homophobic policing in the courts. Others have attempted to remain invisible and have even chased these volunteers away in an effort to protect non-normative sexual practices from governmental intervention. The coalition which mobilized a public case for the exclusion of men from McIvers only engaged with the wider horizon of the city because they were forced to by a complaint of discrimination which threatened the women-only status of the pool. Until the point of the complaint, they mostly kept quiet. The question of whether or not to mobilize some concept of 'the city' as a shared horizon and object of concern is not a given, then, but rather a matter of debate and pragmatic calculation for those struggling to sustain different scenes of public address and circulation. Connections with 'strangers' are sought in the process of these struggles. However, these connections are not made on the basis of shared strangerhood, but through the very labour which is required to sustain a public sphere with *particular* strangers (Ahmed 2000; Iveson 2006).

In concluding, then, I am proposing a vision of the public city as an emergent and contingent product of political labours which are conducted *on behalf of particular forms of publicness*, like protest, and graffiti, and cruising, and women-only swimming, and hanging out in streets at night. The public city, as I imagine it, does not emerge from a universally shared commitment to a normative ideal of the 'being-together of strangers' which stands outside of such purposive public action. In suggesting these revisions to our imagination of the public city, my point is not to suggest that a concern with the intrinsic dimensions of public action and its relationship to 'the city' is entirely misplaced. After all, this is a book which claims to be about 'publics and the city'! Rather, my point is that those of 'us' – academics, students, planners, activists and others – whose primary concern is with the nature of urban life should recognize that this concern, like the concerns of others, is particular to 'our' public, and that our efforts to pursue this concern are therefore purposive and instrumental too. As such, the political utility of attempts to theorize the 'public city' rests on our capacity to establish connections between our own concern for 'the city' and the concerns of others who are engaged in struggles over the making of a variety of publics. My hope is that the analysis presented in this book points to one way in which we might mobilize some concept of 'the city' to reflect on the connections between different struggles to make publics, while not neglecting the particular *things* that are at stake in these struggles (Latour 2005). If in 1994 I began to think that my own concerns about the policing of young people from non-English-speaking backgrounds in Sydney had something to do with 'the city', I am now anxious not to lose sight of the fact that our efforts at YAPA emerged from a quite particular desire to support kids who were angry about being hassled by cops in pedestrian malls while hanging out with friends.

Notes

Chapter 1

1 On shopping malls, see for example Crawford (1992), Goss (1993). On gated communities, see for example Blakely and Snyder (1997) and Low (2003). On quality of life ordinances, see for example Mitchell (2003), Smith (1998). On surveillance, see for example Fyfe and Bannister (1996). On the dominance of the car, see for example Engwicht (1999) and Newman and Kenworthy (1992).

2 Although of course there are significant differences between these two urban administrations, their approach to the question of public space shares some important features. The conservative rhetoric and zero tolerance approach to law and order pioneered by New York Mayor Giuliani has been well documented (see for example Smith 1998). In Barcelona, one of the planners working for Mayor Maragall once noted with regard to public space improvements: 'We . . . thought we were doing a good thing by making a genuine living room in the city with the use of concentric benches. The square as a living room has become such an enormous success that it is mainly used by marginals. Every day they do things there we'd rather not have to see' (quoted in McNeill 1999: 150).

3 Here, concerns about the loss of urban public space match the concerns of other analysts with the loss of public debate and dialogue that is supposed to have been associated with the rise of corporate mass media.

4 Alternatively, a computer with a webcam and internet access might be deployed if one wanted to be *seen* from such a place. I take up such issues in further detail in Chapter 2.

5 For the time being and for the purposes of argument, I will leave the definite article before 'public' in place – the further complications arising from the notion that there are multiple publics are addressed further in Chapter 2.

6 The inverse is also true – that is, we ought to be attuned to the consequences of different forms of public address for the production of different kinds of space. I take this point up in the following chapter.

7   The case studies in Chapters 3–7 each explore the ways in which different contexts for action are mobilized in the creation of the publics under examination.

8   Here, Weintraub (1997: 5) also notes that 'This individual/collective distinction can, by extension, take the form of a distinction between part and whole (of some social collectivity)'.

Chapter 2

1   Indeed, this horizon of strangers is also a condition of 'private address', which seeks (not always successfully) to reach a particular person (or people) directly while avoiding circulation to strangers. I will return to the concept of 'strangers' later in this chapter.

2   'To address a public, we don't go around saying the same thing to all these people. We say it in a venue of indefinite address and hope that people will find themselves in it' (Warner 2002: 86).

3   It is worth noting here that 'oppositional' counterpublics are not by definition progressive, nor is their subordination based on ascribed (rather than elected) characteristics. Fraser might well have added extremist nationalist political groups to her list of groups who have found it advantageous to constitute alternative publics.

4   Indeed, Warner (2002: 119) argues that this risk in fact defines a counterpublic, which by its nature 'maintains at some level, conscious or not, an awareness of its subordinate status'.

5   Both Young (2000: 213–214) and Mitchell (2003: 141) have argued that the importance of urban 'public spaces' for the public sphere lies in the 'embodied' (Young) or 'less mediated' (Mitchell) forms of interaction that they sustain.

6   Indeed, the importance of mediatization suggests that there is a need for scholars to overcome the rather unfortunate tendency for studies of the public sphere to focus on either the city or the media as venues of public address and deliberation. The accessibility and norms of different forms of media are clearly significant in determining the scope of debates about sites in the city. Similarly, various positions in debates about media such as the internet are often informed by urban imaginaries which often go unexamined (Crang 2000). This book goes part of the way here; there is plenty of scope for further work on this problem.

7   The structure of my argument here is similar to the arguments of Castree (2004) in relation to the use of the terms 'culture' and 'economy' in economic geography. As he notes:

> There *are* real and consequential things, relations and processes that the terms culture and economy refer to . . . [W]hether we retain the vocabulary of 'economy' and 'culture' *or* replace it with non-dualistic neologisms, the ontological referents of these terms do not speak for themselves. That is, these terms are not simply mimetic in the work that they do. This leads, second, to the argument that economy and culture need to be seen as two powerful ideas whose coninued use (or eclipse) has important consequences within and beyond geography (209).

Ironically, perhaps, Castree's argument here is developed in a critique of Don Mitchell's claim that there is 'no such thing as culture'. He argues that:

> Mitchell gives short shrift to the possibility that non-dominant groups can actively rework key concepts for counter-hegemonic purposes. But surely the strategic appropriation of 'commonsense' ideas – including but extending beyond the idea of culture – can be a key element of subaltern strategies (213).

This, it seems to me, is exactly Mitchell's position in relation to the idea of 'public space'.

8   Yes, the Men in Kilts New York exist! See http://www.mikny.com/ (last accessed May 2006).

9   'The city', here, may be imagined in relation to other imagined communities such as the nation, the state, the region or the neighbourhood.

Chapter 3

1   In 2002, in a state of post-September 11 security anxiety, the Howard Government closed access to the grassed ramps to enable the construction of a permanent 'terror-proof' wall around the Parliament at a cost of around AUS$22 million. The wall is intended to prevent vehicles from accessing the roof of the building, thus protecting the Parliament from the threat of vehicle-delivered suicide bombs. Secured access to the grassed slopes for pedestrians will be re-established upon completion of the wall. Giurgola himself has been involved in the design and construction of the wall, noting that although he would rather it not be constructed at all, 'In the end, I thought it was better for me to put my hand over it than leave it to someone else. It was my duty' (quoted in *Sydney Morning Herald*, 16–17 April 2005, Spectrum: 7).

2   Section 12(2)(c) of this Act states that: 'A person who being in or on Commonwealth premises, refuses or neglects, without reasonable excuse, to leave these premises on being directed to do so by a constable, by a protective service officer, or by a person authorized in writing by a Minister or the public authority under the Commonwealth occupying the premises to give directions for the purposes of this section, is guilty of an offence.'

3   The regulation of protest at Parliament House is complicated by the fact that not all of Federation Mall (the large grassed mall between Parliament House and Old Parliament House) is a part of the Parliamentary Precincts controlled by the Presiding Officers. Some way down the mall, at the boundary of the Parliamentary Precincts, administrative responsibility shifts to the National Capital Authority (NCA), responsible for the management of National Land, including the Parliamentary Triangle of Canberra. Here, the normal criminal law applies to the conduct of protest, although the NCA has power under Section 8 of the *Trespass on Commonwealth Lands Ordinance* to authorize or remove (with the assistance of the AFP) any structures which protestors might seek to erect or occupy.

4   The Inquiry also considered protests on national land outside foreign embassies, but this will not be discussed here.

5   'Beef' here is a colloquial term for 'complaint' or 'grievance'.

6   The 'sea of hands' was declared a 'structure' for planning purposes (!), and as such the NCA refused the organizers permission to leave the installation overnight – leading to a mad rush of activist activity to install, document and remove the many thousands of hands in daylight hours.

Chapter 4

1   I return to the figure of the 'stranger' in concluding this chapter. At this point, suffice it to say that men who cruise are addressing not just any stranger, but particular strangers – as noted in Chapter 2, this has implications for how we understand the urban dimensions of counterpublic-making.

2   It is important to recognize that only certain configurations of heterosexuality are normalized, and that these configurations are historically and geographically differentiated. So, for example, heterosexual porn or sado-masochism is marginalized and subject to repressive regulatory interventions in various contexts (Warner 2000).

3   Bech (1997: 99, 100) argues that the 'One cannot be homosexual . . . without feeling potentially monitoried', and thus vigilance is combined with an association of desire with 'secrecy, danger and the police'.

4   In other jurisdictions, attempts to deal with the transmission of HIV/AIDS were not modelled on harm-minimization but rather sought to curtail opportunities for homosexual sex, through the closing of saunas and strict policing of beats. In some instances, such measures have been supported by gay and lesbian organizations on the grounds that public sex venues give these communities a bad reputation and are an inherent health risk (see Colter, Hoffman et al. 1996).

5   Indeed, one (male) elected committee member resigned from the PLGLC in 1992, arguing that it focused too much on beats: 'The Committee has . . . failed to ensure continuing representation from the lesbian community, meaning that it has focused exclusively on male issues' (*Brother Sister*, 15 May 1992: 5).

6   The case is unreported, and the name of the defendant was suppressed by the magistrate.

7   The *7.30 Report* is nightly current affairs programme screened nationally.

8   In 1994, a police raid on the Tasty Nightclub, a gay and lesbian event at the Commercial Hotel in the centre of Melbourne, came to symbolize for many the ongoing homophobia of the Victoria Police. At 2.15 a.m. the music stopped, the lights came on, and over 40 police entered the club. No one was allowed to leave. Over the next three hours, 463 patrons and staff were strip-searched. The raid netted a grand total of six arrests on minor drugs possession charges. Outrage ensued, first from members of the lesbian and gay community, and then from others in the wider public sphere. The then Victorian Premier, Jeff Kennett, declared himself disturbed by the raid (*The*

Age, 22 August 1994: 13). During the next few years, over 240 individuals took legal action against the Victoria Police, winning several millions of dollars in compensation (*The Age,* 29 November 1997: Good Weekend, 16–19).

9 Note that Patton positions the sexual vernacular as a 'private language' – while I think her use of the term 'private' here is incorrect, it does signal the important point that participants in counterpublics actively seek to manage the 'counterpublic horizon' given the risks of their participation – as discussed in Chapter 2.

10 While I have argued that beat sex has never been a totally individualistic or privatized action – there is counterpublic communication on the beat – Duncan's point still stands.

Chapter 5

1 Some of Mayor Koch's press statements and advertisements are featured in the 1983 PBS documentary *Style Wars,* directed by Tony Silver. *Style Wars* was reissued on DVD by Plexifilm in 2003. It is almost worth the price of the DVD just to see actors from the cast of the TV series *Fame* trying to convince young people that graffiti is uncool while sitting in a subway car wearing leotards.

2 Tim Cresswell (1996: 51) argued that the authorized placement of graffiti compromised the transgressive nature of graffiti by giving it a 'proper place'. In setting up this opposition between graffiti in the gallery and graffiti on the street/subway, he failed to recognize the ways in which gallery shows could help to circulate the graffiti *as a form* to a wider audience. That is to say, he saw in the gallery only the actual texts on the wall, rather than exploring the interaction between these texts and other graffiti texts in other places.

3 Patrick Neate (2004: 131) also notes the importance of the Buffalo Gals video clip for the making of hip hop cultures in South Africa in his book *Where You're At: Notes from the Frontline of a Hip Hop Planet.*

4 The number of tags supposedly required seems very arbitrary, although it came to be accepted as truth in the wider public sphere and was in due course repeated in other venues, even the Parliament during debates about new graffiti penalties in 1994 (see NSW Parliamentary Debates, Legislative Assembly, Third Series, Vol. 244, 1994: 5976). And the authors also seem a little confused about the nature of a writer's tag, which is not the same as their initials. Their attribution of the male gender to graffiti writers, however, seems to reflect the dominance of young men in graffiti-writing crews, according to Ian Maxwell (1997: 55), who conducted field work in the Sydney hip hop scene in the early 1990s.

5 Perhaps Carr derived this image from the 'City of Fear' articles – pictured next to one of the articles was a gang of dangerous youths, wearing gang colours and some with the ubiquitous baseball caps. The fine print noted that the picture was not in fact of young people from Sydney, but a still-frame taken from the 1988 movie *Colors,* set in Los Angeles.

6 It has been very difficult to establish why the GTF was disbanded by Labor. Nevertheless, the NSW Police continue to gather intelligence and target

graffiti writers in a manner established by the GTF, launching similar 'sting' operations in various train yards (for example the arrest of 14 writers in a sting operation, reported in *Daily Telegraph*, 6 March 2000: 7).

7 See for example *NSW Legislative Council Hansard*, 7 March 2006: 21134, 8 March 2006: 21221, and 6 April 2006: 22265.

8 I was also employed by MYRC as a support worker on this project for six months in 1995.

9 Spanos is not the only property-owner in inner-city Sydney who has been denied permission to allow graffiti artists to paint their walls. In another instance, heritage regulations were invoked to force a residential property owner in Surry Hills to remove a graffiti mural they had commissioned for a wall on their property (see *Sydney Morning Herald*, 13 December 2002: 7).

10 This conference was co-sponsored by Keep Australia Beautiful, the group usually associated with anti-litter campaigns. Graffiti has often been likened to garbage or rubbish in its effects in cities (Cresswell 1996: Ch 3).

11 Similarly, the 2006 Commonwealth Games in Melbourne were preceded by a 'graffiti blitz' (*The Age*, 2 March 2006: 4)

12 Of course, the fact that such a comment was screened on national television also signals some kind of engagement with 'mainstream society', even if it was a hostile one.

13 The Wooster Collective graffiti blog was blocked in 2005 by Manchester Public Library (see http://www.woostercollective.com/2005/06/12-week/, accessed 29 January 2006).

14 In the current climate, the resonance of these images with images of terrorists is hard to ignore. Graffiti writers have indeed been positioned recently through the lens of the 'war on terror' by some politicians, who argue that if graffiti writers are still able to gain access to the Sydney rail network, then it must also be vulnerable to terrorist attack (*Sydney Morning Herald*, 27 December 2005: 3).

Chapter 6

1 According to both studies, this was not a consequence of journalistic bias against Aborigines. Rather, the large number of articles directly referring to Aboriginality in the context of juvenile crime reflected the language used by sources for crime news. The single largest source for such articles was the police (Sercombe 1995: 84).

2 Of course, this concept of an 'adult entertainment precinct' does not extend to the kinds of products on offer at the Club X 'Adult Shop' which Lord Mayor Natrass had so vigorously opposed!

3 Posted at http://abc.net.au/news/australia/wa.

4 This poem is reproduced here with its original punctuation and spelling.

5 In its published form, the poem's power is accentuated by its layout and presentation: with its bare typeface and phonetic spelling, you can almost feel VOID slamming down on the keys as he/she types.

6   Indeed, Gallop and Blair are good mates, having attended university together and been 'best man' at each other's weddings (*Sydney Morning Herald*, 17 January 2006: 1).

Chapter 7

1   Kingsford is a neighbouring suburb to Coogee in Sydney's east.
2   When the *ADA* was amended in the early 1990s to cover discrimination on the grounds of race, transgender and age, the new legislation made reference to both 'the provision of goods and services' *and* 'access to a place', thereby introducing a conceptual distinction which did not previously exist. The Parliament did not subsequently amend existing grounds of discrimination covered by the ADA (such as gender and homosexuality) to include 'access to a place'. As a consequence, discrimination in relation to 'access to a place' on the grounds of sexuality or gender is not covered by the *ADA*.
3   The Bondi Icebergs club takes its name from the membership requirements of the club: one must swim in Bondi Beach's ocean pool on at least three Sundays every winter month from May to October for five years running. At the start of the competitive swimming season in May, competitors throw large blocks of ice into the water to make sure that it is cold enough.
4   When McIvers was built, Council By-laws made it illegal for any bathing in public view during daylight hours. When this restriction was relaxed in 1902, mixed bathing in public view remained illegal for some years.
5   This makes for an interesting contrast with those graffiti writers considered in Chapter 5 who sought to establish their form of public address by establishing graffiti writers as a separate camp which made use of shared space.

Chapter 8

1   This is *not* to say that the place of the audience does not matter. Colomina (1996: 8) goes on to note that in the case of mediatized public events, 'the fact that (for the most part) this audience is indeed at home is not without consequence. The private is, in this sense, now more public than the public'.
2   In 2006, as part of a national protest campaign organized by the Australian Council of Trade Unions to contest a new round of industrial relations 'reform', simultaneous protests were staged in a number of cities, with speeches by union leaders in Melbourne and pre-prepared footage of other union struggles fed 'live' by satellite to giant screens before which protestors gathered.
3   Although, as discussed in the previous section and as observed by others (Barnett 2004), being seen can also be a function of mediatization.
4   It should be clear from this passage that Foucault did not equate 'the eye of power' with a single authority attempting to control the population, but rather with a panoptic mode of governance where the 'eye of power' is internalized by subjects by virtue of being exposed to the vision of others. The 'eye of

power', then, might be the prison warden, or the 'eyes of the street' which Jane Jacobs (1961) famously believed acted to inculcate civility in urban neighbourhoods.

5   Nancy Fraser (2003) notes the similarities between Foucault's analysis of 'the norm' and Arendt's analysis of 'the social'.

6   New York-based billboard bandit and artist Ron English and his collaborators always work during the day: 'If we were spotted at night there would be no doubt that we were up to no good. During the day we were easily spotted and easily taken for guys doing their job' (English 2004: Ch 3).

7   De Certeau's distinction between 'tactics' and 'strategies' as spatial operations is discussed in Chapter 2.

8   I do not mean to suggest that there are not some significant differences among this group of theorists. Rather, my point is that each in their own way comes to the conclusion that urban life 'demands' an 'open' or 'reasonable' disposition towards strangers. For a further discussion, see Iveson (2006a: 79).

References

Ahmed, S. (2000) *Strange Encounters: Embodied Others in Post-Coloniality*: London, Routledge.

Alexander, I. and S. Houghton (1994) 'New Investment in Urban Public Transport' in *Australian Planner*, 32(1): 7–11.

Alexander, I. and S. Houghton (1995) 'New Investment in Urban Public Transport II' in *Australian Planner*, 32(2): 82–87.

Amin, A. and N. Thrift (2002) *Cities: Reimagining the Urban*: Oxford, Polity Press.

Anderson, B. (1983) *Imagined Communities: Reflections on the Origin and Spread of Nationalism*: London, Verso.

Appadurai, A. (1990) 'Disjuncture and Difference in the Global Cultural Economy' in *Public Culture*, 2(2): 1–24.

Arendt, H. (1958) *The Human Condition*: Chicago, University of Chicago Press.

Auslander, P. (1999) *Liveness: Performance in a Mediatized Culture*: London and New York, Routledge.

Austin, J. (2001) *Taking the Train: How Graffiti became an Urban Crisis in New York City*: New York, Columbia University Press.

Australian Heritage Commission (1995) 'Aboriginal Embassy Site Citation' in *Joint Standing Committee on the National Capital and External Territories – Inquiry into the right to legitimately protest or demonstrate on National Land and in the Parliamentary Zone in particular*, 1 – Submissions: 197–9.

Australian Institute of Urban Studies WA Branch (Ed.) (1992) *Capital City Planning for Perth, the Capital of Western Australia*: Perth, Australian Institute of Urban Studies.

Australian Labor Party (NSW) (1994) *Law and Order Policy*: Sydney, Australian Labor Party (NSW).

Baird, B. (1997) 'Putting police on notice' in S.A. Tomsen and G. Mason (Eds.) *Homophobic Violence*: Sydney, Hawkins Press.

Barnett, C. (2004) 'Media, Democracy and Representation: Disembodying the Public' in C. Barnett and M. Low (Eds.) *Spaces of Democracy: Geographical Perspectives on Citizenship, Participation and Representation*: London, Sage.

Barnett, C. (forthcoming) 'Convening Publics: The parasitical spaces of public action' in K. Cox, M. Low and J. Robinson (Eds.) *The Handbook of Political Geography*: London, Sage.

Bartos, M. (1996) 'The queer excess of public health policy' in *Meanjin*, 55(1): 122–31.

Bartos, M., J. McLeod and P. Nott (1993) *Meanings of Sex Between Men*: Canberra, Australian Government Publishing Service.

Bech, H. (1997) *When Men Meet Men: Homosexuality and Modernity*: Chicago, University of Chicago Press.

Bech, H. (1998) 'Citysex: representing lust in public' in *Theory, Culture and Society*, 15(3–4): 215–41.

Benhabib, S. (1992) 'Models of Public Space: Hannah Arendt, the Liberal Tradition, and Jurgen Habermas' in C. Calhoun (Ed.) *Habermas and the Public Sphere*: Cambridge, MA, MIT Press.

Benhabib, S. (2002) *The Claims of Culture: Equality and Diversity in the Global Era*: Princeton, Princeton University Press.

Benn, S.I. and G.F. Gaus (1983) 'The public and the private: concepts and action' in S.I. Benn and G.F. Gaus (Eds.) *Public and Private in Social Life*: London and Canberra, Croom Helm.

Berlant, L. and M. Warner (1998) 'Sex in Public' in *Critical Inquiry*, 24(2): 547–66.

Blagg, H. (2005) *Background paper on: A New Way of Doing Justice Business? Community Justice Mechanisms and Sustainable Governance in Western Australia*: Perth, Law Reform Commission of Western Australia, Background Paper Number 8.

Blagg, H. and G. Valuri (2004) 'Self-policing and community safety: the work of Aboriginal Community Patrols in Australia' in *Current Issues in Criminal Justice*, 15(3): 205–19.

Blagg, H. and M. Wilkie (1995) *Young People and Police Powers*: Sydney, Australian Youth Foundation.

Blair, T. (2001) *Improving Your Local Environment*, Speech to Groundwork Trust, Croydon, UK. Available at http://www.number-10.gov.uk/output/Page1588.asp. Accessed 9 January 2006.

Blakely, E.J. and M.G. Snyder (1997) *Fortress America: Gated Communities in the United States*: Washington, Brookings Institution Press.

Boddy, T. (1992) 'Underground and Overhead: Building the Analogous City' in M. Sorkin (Ed.) *Variations on a Theme Park – The New American City and the End of Public Space*: New York, Hill and Wang.

Brenner, N. and N. Theodore (2005) 'Commentary: Neoliberalism and the Urban Condition' in *City*, 9(1): 101–8.

Brown, M. (1997) *RePlacing Citizenship: AIDS Activism and Radical Democracy*: New York, Guildford Press.

Brown, M. (1999) 'Reconceptualizing Public and Private in Urban Regime Theory: Governance in AIDS Politics' in *International Journal of Urban and Regional Research*, 23(1): 70–87.

Butsch, R. (2000) *The Making of American Audiences: From Stage to Television, 1750–1990*: Cambridge, Cambridge University Press.

Calhoun, C. (1996) 'Social Theory and the Public Sphere' in B.S. Turner (Ed.) *A Companion to Social Theory*: Malden, Blackwell.

Calhoun, C. (1997) 'Nationalism and the Public Sphere' in J. Weintraub and K. Kumar (Eds.) *Public and Private in Thought and Practice: Perspectives on a Grand Dichotomy*: Chicago, University of Chicago Press.

Calhoun, C. (1998) 'Community without Propinquity Revisited: Communications Technology and the Transformation of the Urban Public Sphere' in *Sociological Inquiry*, 68(3): 373–97.

Carbery, G. (1992) 'Some Melbourne Beats: A "Map" of a Subculture' in R. Aldrich and G. Wotherspoon (Eds.) *Gay Perspectives*: Sydney, Department of Economic History, University of Sydney.

Carpignano, P. (1999) 'The Shape of the Sphere: The Public Sphere and the Materiality of Communication' in *Constellations*, 6(2): 177–89.

Carr, D. and R. Fardon (1993) *A Capital City for Western Australia*: Perth, Government of Western Australia.

Carr, S., M. Francis, L.G. Rivlin and A.M. Stone (1992) *Public Space*: London, Cambridge University Press.

Castells, M. (1979) *The Urban Question*: Cambridge, MA, MIT Press.

Castells, M. (1983) *The City and the Grassroots*: Berkeley, University of California Press.

Castleman, C. (1982) *Getting Up: Subway graffiti in New York*: Cambridge, MA, MIT Press.

Castree, N. (2004) 'Economy and culture are dead! Long live economy and culture!' in *Progress in Human Geography*, 28(2), 204–26.

Certeau, M. de (1984) *The Practice of Everyday Life*: Berkeley, University of California Press.

Chalfant, H. and J. Prigoff (1987) *Spraycan Art*: New York, Thames and Hudson.

Chauncey, G. (1994) *Gay New York: Gender, Urban Culture, and the Makings of the Gay Male World, 1890–1940*: New York, Routledge.

Chesluck, B. (2004) ' "Visible Signs of a City Out of Control": Community Policing in New York City' in *Cultural Anthropology*, 19(2): 250–75.

City of Perth (1995a) *Furniture Strategy Draft*: Perth, Perth – A City for People.

City of Perth (1995b) *Lighting Strategy Draft*: Perth, Perth – A City for People.

City of Perth (1995c) *Perth – A City for People: Strategic Management Plan, 1995–2000*: Perth, City of Perth.

City of Perth (1997) *Youth Forum: Report of Findings*: Perth, City of Perth.

City of Perth (1999) *Safety and Security Action Plan (Draft)*: Perth, City of Perth.

City of Perth (nd) *Closed Circuit TV Surveillance System Information Kit*: Perth, City of Perth.

Citysafe (1996) *Perth the Vision of a Safe City: The Role of Community Policing in the City of Perth*: Perth, City of Perth and WA Police.

CityVision (1992 (1988)) 'New Directions for Central Perth' in Australian Institute of Urban Studies (Ed.) *Capital City Planning for Perth, the Capital of Western Australia*: Perth, Australian Institute of Urban Studies.

Classification Review Board (2006) *Mark Ecko's Getting Up: Contents Under Pressure*: Minutes of meetings held on February 6, 8, 13 and 14 2006, available at http://www.oflc.gov.au/resource.html?resource=794&filename=794.pdf, accessed 17 May 2006.

Cloughley, G. (2003) *Lament,* Transcript of talk on ABC Radio National 14 May 2003. Available at http://www.abc.net.au/rn/talks/perspective/stories/s854354.htm, accessed 21 April 2005.

Cohen, S. (1985) *Visions of Social Control: Crime, Punishment and Classification:* Oxford, Blackwell.

Collins, A. (1998) 'Hip Hop Graffiti Culture' in *Alternative Law Journal,* 23(1): 19–21.

Collins, A.M. (1997) 'Hip Hop Graffiti Culture (HHGC): Addressing Social and Cultural Aspects' in *Graffiti: Off the Wall Seminar Proceedings, University of New South Wales, 11 June:* Sydney, http://www.graffiti.nsw.gov.au.

Collins, D. and R. Kearns (2001) 'Under Curfew and Under Siege? Legal Geographies of Young People' in *Geoforum,* 32(3): 389–403.

Colomina, B. (1996) *Privacy and Publicity: Modern Architecture as Mass Media:* Cambridge, MA, MIT Press.

Colter, E.G., W. Hoffman, E. Pendelton, A. Redick and D. Serlin (Eds.) (1996) *Policing Public Sex : Queer Politics and the Future of AIDS Activism:* Boston, MA, South End Press.

Commonwealth of Australia (1993) *Expressing Australia: Art in Parliament House:* Canberra, Joint House Department of Parliament House, Canberra.

Cooper, A., R. Law, J. Malthus and P. Wood (2000) 'Rooms of their Own: Public Toilets and Gendered Citizens in a New Zealand City, 1860–1940' in *Gender, Place and Culture,* 7(4): 417–33.

Cooper, D. (1998) 'Regard between Strangers: Diversity, Equality and the Reconstruction of Public Space' in *Critical Social Policy,* 18(4): 465–92.

Cooper, M. and H. Chalfant (1984) *Subway Art:* London, Thames and Hudson Ltd.

Coxon, A.P. (1996) *Public Sex Environments: Methods and Findings:* Essex, Project Sigma, Working Paper 5.

Crang, M. (2000) 'Public Space, Urban Space and Electronic Space: Would the Real City Please Stand Up?' in *Urban Studies,* 37(2): 301–17.

Crawford, E. (2003) 'Equity and the City: The Case of the East Perth Redevelopment' in *Urban Policy and Research,* 21(1): 81–92.

Crawford, M. (1992) 'The World in a Shopping Mall' in M. Sorkin (Ed.) *Variations on a Theme Park – The New American City and the End of Public Space:* New York, Hill and Wang.

Cresswell, T. (1996) *In Place/Out of Place: Geography, Ideology, and Transgression:* Minneapolis, University of Minnesota Press.

Cresswell, T. (1998) 'Night Discourse: Producing/Consuming Meaning on the Street' in N.R. Fyfe (Ed.) *Images of the Street: Planning, Identity and Control in Public Space:* London, Routledge.

Curtiss, K. (1998) *Graffiti Solutions Local Government Pilot Projects – Hurstville City Council Final Report:* Sydney, Hurstville City Council, see also http://www.graffiti.nsw.gov.au.

da Landa, M. (1986) 'Policing the Spectrum' in *Zone 1/2: City,* 1/2: 177–87.

Davis, M. (1990) *City of Quartz: Excavating the Future in Los Angeles:* London, Verso.

Davis, M. (1992) 'Fortress Los Angeles: The Militarization of Urban Space' in M. Sorkin (Ed.) *Variations on a Theme Park – The New American City and the End of Public Space:* New York, Hill and Wang.

Davison, G. (1994) 'Public Life and Public Space: A Lament for Melbourne's City Square' in *Historic Environment*, 11(1): 4–9.

Dean, M. (2002) 'Liberal Government and Authoritarianism' in *Economy and Society*, 31(1): 37–61.

Delph, E.W. (1978) *The Silent Community: Public Homosexual Encounters*: Beverly Hills, Sage.

Department of Planning and Urban Development (WA) (1990) *Metroplan: A Planning Strategy for the Perth Metropolitan Region*: Perth, Government of Western Australia.

Department of Planning and Urban Development (WA) (1993) *Metropolitan Attitudes Survey June 1993*: Perth, Inner City Living.

Department of Planning and Urban Development (WA) (1994) *Intending Residents Register Summary*: Perth, Inner City Living.

Deutsche, R. (1996) *Evictions: Art and Spatial Politics*: Cambridge, MA, MIT Press.

Deutsche, R. (1999) 'Reasonable Urbanism' in J. Copjec and M. Sorkin (Eds.) *Giving Ground: The Politics of Propinquity*: London, Verso.

Donald, J. (1999) *Imagining the Modern City*: London, Athlone.

Dorrian, M., D. Recchia and L. Farrelly (2002) *Stick 'Em Up!*: London, Booth–Clibbord Editions.

Dowsett, G.W. (1996) *Practicing Desire: Homosexual Sex in the Era of AIDS*: Stanford, Stanford University Press.

Dowsett, G.W. and M.D. Davis (1992) *Transgression and Intervention: Homosexually Active Men and Beats*: Sydney, National Centre for HIV Social Research, Macquarie University.

d'Souza, M. and K. Iveson (1999) 'Homies and Homebrewz: Hip Hop in Sydney' in R. White (Ed.) *Australian Youth Subcultures: On the Margins and in the Mainstream*: Hobart, Australian Clearinghouse for Youth Studies.

Duncan, N. (1996) 'Negotiating Gender and Sexuality in Public and Private Spaces' in N. Duncan (Ed.) *Bodyspace: Destabilizing Geographies of Gender and Sexuality*: London, Routledge.

Eley, G. (1992) 'Nations, Publics and Political Cultures: Placing Habermas in the Nineteenth Century' in C. Calhoun (Ed.) *Habermas and the Public Sphere*: Cambridge, MA, MIT Press.

Emirbayer, M. and M. Sheller (1999) 'Publics in History' in *Theory and Society*, 28(1): 145–97.

English, R. (2004) *Popaganda*: New York, Last Gasp (Also available online at www.popaganda.com).

Engwicht, D. (1999) *Street Reclaiming: Creating Liveable Streets and Vibrant Communities*: Sydney, Pluto Press.

Fainstein, S.F. (1999) 'Can we make the cities we want?' in R.A. Beauregard and S. Body-Gendrot (Eds.) *The Urban Moment: Cosmopolitan Essays on the Late-20th-Century City*: Thousand Oakes, Sage.

Featherstone, M. (1998) 'The Flaneur, the City and Virtual Public Life' in *Urban Studies*, 35(5–6): 909–25.

Fincher, R. and R. Panelli (2001) 'Making Space: Women's Urban and Rural Activism and the Australian State' in *Gender, Place and Culture*, 8(2): 129–48.

Foucault, M. (1977) *Discipline and Punish: The Birth of the Prison*: London, Allen Lane.

Foucault, M. (1980) 'The Eye of Power' in C. Gordon (Ed.) *Power/Knowledge – Selected Interviews and Other Writings, 1972–1977*: Brighton, Harvester Press.

Foulkes, D. (1992) 'An Application for Exemption from the Provisions of the Equal Opportunity Act 1984 (Vic.) by the City of Brunswick ("The *Brunswick Baths Case*")' in *Melbourne University Law Review*, 18(June): 711–15.

Fraser, D. (1995) 'Father Knows Best: Transgressive Sexualities (?) and the Rule of Law' in *Current Issues in Criminal Justice*, 7(1): 82–7.

Fraser, N. (1992) 'Rethinking the Public Sphere: A Contribution to the Critique of Actually Existing Democracy' in C. Calhoun (Ed.) *Habermas and the Public Sphere*: Cambridge, MA, MIT Press.

Fraser, N. (2003) 'From Discipline to Flexibilization? Rereading Foucault in the Shadow of Globalization' in *Constellations*, 10(2): 160–71.

Fyfe, N.R. (1996) 'City Watching: Closed Circuit Television Surveillance in Public Spaces' in *Area*, 28(1): 37–46.

Gallop, G. (2003a) *Premier Unveils Northbridge Curfew Policy*: Perth, Premier's Office, Government of Western Australia.

Gallop, G. (2003b) *Speech delivered at the launch of the Rio Tinto Child Health Partnership*: Telethon Institute for Child Health Research, Available at http://www.premier.wa.gov.au/docs/speeches/RioTintoChildHealth_250903_FINAL.pdf, accessed May 2006.

Gaonkar, D.P. and E. Povinelli (2003) 'Technologies of Public Forms: Circulation, Transfiguration, Recognition' in *Public Culture*, 15(3): 385–97.

Geason, S. and P. Wilson (1990) *Preventing Graffiti and Vandalism*: Canberra, Australian Institute of Criminology.

Gehl, J. (1992) 'Giving the City a Human Face' in *Perth Beyond 2000: A Challenge for the City – Proceedings from City Challenges Conference*: Perth, Government of Western Australia.

Gehl, J. (1994) *Public Spaces and Public Life in Perth*: Perth, Government of Western Australia and City of Perth.

Goffman, E. (1959) *The Presentation of Self in Everyday Life*: London, Mayflower.

Goss, J. (1993) 'The "Magic of the Mall": An Analysis of Form, Function, and Meaning in the Contemporary Retail Built Environment' in *Annals of the Association of American Geographers*, 83(1): 18–47.

Government of Western Australia (1990) *The East Perth Project: Outline Development Plan*: Perth, East Perth Project Group.

Government of Western Australia (1995) *A Cultural Centre for a Capital City: Draft Report*: Perth, Government of Western Australia.

Government of Western Australia (2003) *Government Achievements Report 24 March – 28 April 2003*: Perth, Government of Western Australia.

Government of Western Australia and City of Perth (1996) *A Strategy for Children in the City of Perth*: Perth, Government of Western Australia.

Government of Western Australia and City of Perth (1997) *Enhancing Perth's Public Places*: Perth, Perth – A City for People.

Habermas, J. (1989 [1962]) *The Structural Transformation of the Public Sphere*: Cambridge, MA, MIT Press.

Haebicc, A. (1992) *For Their Own Good: Aborigines and Government in the South West of Western Australia, 1900–1940*: Perth, UWA Press.

Halsey, M. and A. Young (2002) 'The Meanings of Graffiti and Municipal Administration' in *The Australian and New Zealand Journal of Criminology*, 35(2): 165–86.

Handley, R. (1987) 'Keeping People Away from Parliament' in *Legal Service Bulletin*, 12(1): 25–7.

Harvey, D. (1997) *Justice, Nature and the Geography of Difference*: Cambridge, MA, Blackwell.

Harvey, D. (2000) *Spaces of Hope*: Edinburgh, University of Edinburgh Press.

Healey, K. (Ed.) (1996) *Youth Gangs*: Sydney, Spinney Press.

Healey, P. (2002) 'On Creating the "City" as a Collective Resource' in *Urban Studies*, 39(10): 1777–92.

Hebdige, D. (1988) *Hiding in the Light: On Images and Things*: London, Routledge.

Hickman, P. (1991) 'Transit Crime and Policing' in *Australian Police Journal*, 45(3): 98–102.

Hindess, B. (1996) *Discourses of Power: From Hobbes to Foucault*: Oxford, UK; Cambridge, MA, Blackwell.

Hodge, S. (1995) '"No Fags Out There": Gay Men, Identity and Suburbia' in *Journal of Interdisciplinary Gender Studies*, 1(1): 41–8.

Holston, J. and A. Appadurai (1996) 'Cities and Citizenship' in *Public Culture* (8): 187–204.

Home Office (UK) (2003) *Respect and Responsibility – Taking a Stand Against Anti-Social Behaviour*: London, Home Office.

Howe, B. (1994) *News Release – Parliamentary Committee to Look at Protests on National Land*: Canberra, Brian Howe, Deputy Prime Minister, Minister for Housing and Regional Development.

Hubbard, P. (2002) 'Sex Zones: Intimacy, Citizenship and Public Space' in *Sexualities*, 4(1): 51–71.

Hull, P., P. Van Deven, G. Prestage, P. Rawstone, S. Kippax, G. Horn, M. Kennedy, G. Hussey and C. Batrouney (2004) *Gay Community Periodic Survey: Melbourne 2004*: Sydney, National Centre in HIV Social Research.

Hume, J. (1996) 'Parliament House: Securing our Nation' in *Security Australia*, 17(6): 8–9.

Iveson (2001) 'Counterpublics and Public Space: Comparing Labour Movement and Aboriginal Protest at Parliament House, Canberra' in R. Markey (Ed.) *Labour and Community*: Wollongong, University of Wollongong Press.

Iveson, K. (1998) 'Putting the Public Back into Public Space' in *Urban Policy and Research*, 16(1): 21–33.

Iveson, K. (2003) 'Justifying Exclusion: The Politics of Public Space and the Dispute over Access to McIvers Ladies' Baths, Sydney' in *Gender, Place and Culture*, 10(3): 215–28.

Iveson, K. (2006a) 'Strangers in the Cosmopolis' in J. Binnie, J. Holloway, S. Millington and C. Young (Eds.) *Cosmopolitan Urbanism*: London, Routledge.

Iveson, K. (2006b) 'Cities for angry young people? From exclusion and inclusion to engagement in urban policy' in B.J. Gleeson and N. Sipe (Eds.) *Creating Child Friendly Cities*: London, Routledge.

Jacobs, J. (1961) *The Death and Life of Great American Cities*: New York, Random House.

Jacobs, J.M. (1996) *Edge of Empire: Postcolonialism and the City*: London, Routledge.

Jameson, F. (1988) 'On Negt and Kluge' in *October* (46): 151–78.

Johnson, V. (1997) *Michael Jagamara Nelson*: Sydney, Craftsman House.

Joint Standing Committee on the National Capital and External Territories (1995a) *Inquiry into the right to legitimately protest or demonstrate on National Land and in the Parliamentary Zone in particular: submissions*: Canberra, Parliament of Australia.

Joint Standing Committee on the National Capital and External Territories (1995b) *Inquiry into the right to legitimately protest or demonstrate on National Land and in the Parliamentary Zone in particular: transcript of evidence*: Canberra, Parliament of Australia.

Joint Standing Committee on the National Capital and External Territories (1997) *A Right to Protest*: Canberra, Australian Government Publishing Service.

Jordan, J. (1998) 'The Art of Necessity: The Subversive Imagination of Anti-Road Protest and Reclaim the Streets' in G. McKay (Ed.) *DiY Culture: Party and Protest in Nineties Britain*: London, Verso.

Joseph, S. (1997) *She's My Wife, He's Just Sex*: Sydney, Australian Centre for Independent Journalism, University of Technology.

Katz, P. (1994) *The New Urbanism: Towards an Architecture of Community*: New York, McGraw-Hill.

Katznelson, I. (1992) *Marxism and the City*: Oxford, Clarendon Press.

Kearney, R. (2003) *Strangers, Gods and Monsters: Interpreting Otherness*: London, Routledge.

Keith, M. (2001) 'Knowing the City?' in *City*, 5(1): 107–13.

Keith, M. (2005) *After the Cosmopolitan? Multicultural Cities and the Future of Racism*: London, Routledge.

Kelling, G. and C. Coles (1997) *Fixing Broken Windows: Restoring Order and Reducing Crime in Our Communities*: New York, Simon and Schuster.

Kingsford Legal Centre (1995) McIvers Ladies Baths, Coogee, Letter to Hon EP Pickering MLC, 16 February.

Knopp, L. (1995) 'Sexuality and Urban Space: A Framework for Analysis' in D. Bell and G. Valentine (Eds.) *Mapping Desire: Geographies of Sexualities*: London, Routledge.

Koch, T. (2003) 'Aboriginal Legal Service of Western Australia' in *Indigenous Law Bulletin*, 5(27): 7–8.

KPMG (1995) *Perth – A City for People: Report on Submissions and Public Comment Received Volume 1*: Perth, Capital City Committee, Department of Premier and Cabinet.

Kristeva, J. (1991) *Strangers to Ourselves*: New York, Harvester Wheatsheaf.

Larner, W. (2003) 'Neoliberalism?' in *Environment and Planning D: Society and Space*, 21(5): 509–12.

Latour, B. (2005) 'From Realpolitik to Dingpolitik or How to Make Things Public' in B. Latour and P. Weibel (Eds.) *Making Things Public: Atmospheres of Democracy*: Cambridge, MA, MIT Press, 14–43.

Law, R., A. Cooper, J. Malthus and P. Wood (1999) 'Bodies, Sites and Citizens: The Politics of Public Toilets' in *Proceedings of the Southern Regional Conference of the*

International Geographical Union Commission on Gender (February, Dunedin New Zealand): 53–60.

Lefebvre, H. (1991) *The Production of Space*: Oxford, Blackwell.

Liberal Party (WA) and National Party of Australia (WA) (1993) *Project Perth: A Living and Working Capital*: Perth.

LiPuma, E. and T. Koelble (2005) 'Cultures of Circulation and the Urban Imaginary: Miami as Example and Exemplar' in *Public Culture*, 17(1): 153–79.

London Assembly (2002) *Graffiti in London: Report of the London Assembly Graffiti Investigative Committee*: London, Greater London Authority.

Low, S. (2003) *Behind the Gates: Life, Security, and the Pursuit of Happiness in Fortress America*: New York, Routledge.

MacDonald, N. (2001) *The Graffiti Subculture: Youth, Masculinity, and Identity in London and New York*: Basingstoke, Palgrave.

MacLeod, G. (2002) 'From Urban Entrepreneurialism to a Revanchist City? On the Spatial Injustices of Glasgow's Renaissance' in *Antipode*, 34(3): 602–24.

Manco, T. (2004) *Street Logos*: London, Thames and Hudson.

Mason, G. and S. Tomsen (Eds.) (1997) *Homophobic Violence*: Leichhardt, Federation Press.

Maxwell, I. (1997) 'Hip Hop Aesthetics and the Will to Culture' in *Australian Journal of Anthropology*, 8(1): 50–70.

Maxwell, I. (2003) *Phat Beats, Dope Rhymes: Hip Hop Down Under Comin' Upper*: Middletown, Wesleyan University Press.

McDermott, M.-L. (2005) 'Changing Visions of Baths and Bathers: Desegregating Ocean Baths in Wollongong, Kiama and Gerringong' in *Sporting Traditions*, 22(1): 1–20.

McNeill, D. (1999) *Urban Change and the European Left: Tales from the New Barcelona*: London, Routledge.

McNeill, D. and M. Tewdwr-Jones (2003) 'Architecture, Banal Nationalism and Re-territorialization' in *International Journal of Urban and Regional Research*, 27(3): 738–43.

Melbourne City Council (1997) *Review into Public Toilet Provision: Planning, Development and Environment Committee Report*, Agenda item 5.6, MCC meeting 9 September.

Mickler, S. and A. McHoul (1998) 'Sourcing the Wave: Crime Reporting, Aboriginal Youth and the WA Press, Feb 1991–Jan 1992' in *Media International Australia, Incorporating Culture and Policy*, (86): 122–52.

Mitchell, D. (1995) 'The End of Public Space? People's Park, Definitions of the Public, and Democracy' in *Annals of the Association of American Geographers*, 85(1): 108–33.

Mitchell, D. (1996) 'Political Violence, Order, and the Legal Construction of Public Space: Power and the Public Forum Doctrine' in *Urban Geography*, 17(2): 152–78.

Mitchell, D. (1997) 'The Annihilation of Space by Law: The Roots and Implications of Anti-Homeless Laws in the United States' in *Antipode*, 29(3): 303–35.

Mitchell, D. (2003) *The Right to the City: Social Justice and the Fight for Public Space*: New York, Guildford Press.

Moran, L.J., B. Skeggs, P. Tyrer and K. Corteen (2003) 'The Formation of Fear in Gay Space: The "Straights" Story' in *Capital and Class*, 80: 173–98.

Morris, M. (1992) *Great Moments in Social Climbing: King Kong and the Human Fly*: Sydney, Local Consumption Productions.

Murdoch, W. (1998) ' "Disgusting Doings" and "Putrid Practices": Reporting Homosexual Men's Lives in the Melbourne Truth during the First World War' in *Studies in Australian Culture*, (4): 116–31.

Murphy, P. and S. Watson (1994) 'Social Polarization and Australian Cities' in *International Journal of Urban and Regional Research*, 18(4): 572–590.

Neate, P. (2004) *Where You're At: Notes from the Frontline of a Hip Hop Planet*: London, Bloomsbury.

Negt, O. and A. Kluge (1988) 'Public Sphere and Experience: Selections' in *October*, (46): 60–82.

Negt, O. and A. Kluge (1993) *Public Sphere and Experience: Towards an analysis of the Bourgeois and Proletarian Public Sphere*: Minneapolis, University of Minnesota Press.

New Heart for Perth Society (1969) *New Heart News*: Perth, The Society.

New South Wales Police Service (1995) 'Commissioner's Circular: Offensive Behaviour at "Beats" ' in *Police Service Weekly*, 7(39): 22–3.

Newman, P. and J. Kenworthy (1992) *Winning Back the Cities*: Sydney, Australian Consumers' Association, Pluto Press.

Norington, B. (1998) *Jennie George*: Sydney, Allen and Unwin.

Office of Crime Prevention (Western Australia) (2004) *Responsible Parenting Orders and Responsible Parenting Contracts*: Perth, Department of the Premier and Cabinet, Government of Western Australia.

O'Malley, P. (1998) 'Indigenous Governance' in M. Dean and B. Hindess (Eds.) *Governing Australia: Studies in Contemporary Rationalities of Government*: Melbourne, Cambridge University Press.

O'Neill, P.M. and N. Argent (2005) 'Neoliberalism in Antipodean Spaces and Times: An Introduction to the Special Theme Issue' in *Australian Geographical Studies*, 43(1): 2–8.

O'Sullivan, T. (1995) ' "City of Fear" ' in *Alternative Law Journal*, 20(2): 94–6.

Painter, J. and C. Philo (1995) 'Spaces of Citizenship: An Introduction' in *Political Geography*, 14(2): 107–20.

Parliament House Construction Authority (1979a) *Parliament House Canberra: Conditions for a Two Stage Competition, Volume One*: Canberra, Parliament House Construction Authority.

Parliament House Construction Authority (1979b) *Parliament House Canberra: Conditions for a Two Stage Competition, Volume Two*: Canberra, Parliament House Construction Authority.

Parliament House Construction Authority (1986) *Australia's New Parliament House*: Canberra, Parliament House Construction Authority.

Pateman, C. (1983) 'Feminist Critiques of the Public/Private Dichotomy' in S.I. Benn and G.F. Gaus (Eds.) *Public and Private in Social Life*: London, Croom Helm.

Patton, C. (1991) 'Safe Sex and the Pornographic Vernacular' in Bad Object Choices (Ed.) *How Do I Look? Queer Film and Video*: Seattle, Bay Press.

Peck, J. and A. Tickell (2002) 'Neoliberalizing Space' in *Antipode*, 34(3): 380–404.

Peel, M. (1995) 'The Urban Debate: From "Los Angeles" to the Urban Village' in P. Troy (Ed.) *Australian Cities – Issues, Strategies and Policies for Urban Australia in the 1990s*: Melbourne, Cambridge University Press.

Peet, M. (1998) *Graffiti and Art: Hip Hop Graffiti Culture – Interview with Matthew Peet, Graffiti Artist, Rap Musician and Youth Worker, by Ken Dray, Co-ordinator, NSW Graffiti Solutions Program. Monday 28 October 1998*: Sydney, http://www.graffiti.nsw.gov.au.

Perth Central Area Policies Review Study Team (1993) *Perth Central Area Policies Review: Report for Public Comment*: Perth, State Planning Commission.

Perth Inner City Housing Association (1997a) 'Burt Way Apartments Win Reprieve' in *PICHA News*, (8): 5.

Perth Inner City Housing Association (1997b) 'The Story of the Haig Park Development' in *PICHA News*, (8): 4.

Perth Inner City Living Taskforce (1992) *City Living*: Perth, Commonwealth of Australia, Government of Western Australia, City of Perth.

Phillips, A. (1993) *Democracy and Difference*: University Park (PA), Pennsylvania University Press.

Phillips, P.C. (1992) 'Images of Repossession' in P. Fresham (Ed.) *Public Address: Krzysztof Wodiczko*: Minneapolis, Walker Arts Center.

Pickering, T. (1995) *Media Release, 3 March 'Special Needs' Certification for McIver's Baths*: Sydney, Minister for Energy and Minister for Local Government and Co-operatives.

Pile, S. (1997) 'Introduction: Opposition, Political Identities and Spaces of Resistance' in S. Pile and M. Keith (Eds.) *Geographies of Resistance*: London, Routledge.

Pinnington, M. (1998) *Guildford Youth Project, Graffiti Solutions Pilot Project*: Sydney, Parramatta City Council, see also http://www.graffiti.nsw.gov.au.

Planning Group Pty Ltd and Godfrey Spowers Puddy and Lee (1995) *Inner City Residential Manual*: Perth, Inner City Living.

Podmore, J.A. (2001) 'Lesbians in the Crowd: Gender, Sexuality and Visibility along Montreal's Boul. St-Laurent' in *Gender, Place and Culture*, 8(4): 333–55.

Police Lesbian and Gay Liaison Committee (1994) *Annual Report 1994*: Melbourne, PLGLC.

Police Lesbian and Gay Liaison Committee (Vic) (1996) *Policing Public Place 'Beat' Meeting Behaviour – Recommendations for the Victoria Police*: Melbourne, PLGLC.

Powers, S. (1999) *The Art of Getting Over: Graffiti at the Millennium*: New York, St Martin's Press.

Puplick, C. (1995) Letter to Ms Julie Spies re Ladies Baths at Coogee, 27 January.

Quilley, S. (1997) 'Constructing Manchester's "New Urban Village": Gay Space in the Entrepreneurial City' in G.B. Ingram, A.-M. Bouthillette and Y. Retter (Eds.) *Queers in Space: Communities, Public Places, Sites of Resistance*: Seattle, Bay Press.

Rail Corporation New South Wales (2005) *Annual Report*: Sydney, Rail Corporation New South Wales.

Rancière, J. (1999) *Disagreement: Politics and Philosophy*: Minneapolis, University of Minnesota Press.

Rancière, J. (2001) 'Ten Theses on Politics' in *Theory and Event*, 5(3).

Randwick Municipal Council (1995a) Minutes of Extraordinary meeting of Randwick Council, 18 January.

Randwick Municipal Council (1995b) Section 126A Anti-Discrimination Act Application – McIver's Baths, Application to Hon T Pickering MLC, 21 February.

Robbins, B. (1993) 'Introduction: The Public as Phantom' in B. Robbins (Ed.) *The Phantom Public Sphere*: Minneapolis, University of Minnesota Press.

Robins, K. (1996) *Into the Image: Culture and Politics in the Field of Vision*: London, Routledge.

Robinson, J. (2002) 'Global and World Cities: A View from off the map' in *International Journal of Urban and Regional Research*, 26(3): 531–54.

Rose, N. (2000) 'Government and Control' in *British Journal of Criminology*, 40(2): 321–39.

Ruddick, S. (1996) *Young and Homeless in Hollywood: Mapping Social Identities*: New York, Routledge.

Sandercock, L. (1998) *Towards Cosmopolis: Planning for Multicultural Cities*: Chichester, John Wiley and Sons.

Sandercock, L. (2003) *Cosmopolis II: Mongrel Cities in the 21st Century*: London, Continuum.

Scalmer, S. (2001) *Dissent Events: Protest, the Media, and the Political Gimmick in Australia*: Sydney, University of New South Wales Press.

Schivelbusch, W. (1987) 'The Policing of Street Lighting' in *Yale French Studies*, 73: 61–74.

Sennett, R. (1978) *The Fall of Public Man*: New York, Random House.

Sennett, R. (1990) *The Conscience of the Eye*: London, Faber and Faber.

Sennett, R. (1994) *Flesh and Stone: The Body and the City in Western Civilization*: London, Faber and Faber.

Sercombe, H. (1995) 'The Face of the Criminal is Aboriginal: Representations of Aboriginal Young People in the West Australian newspaper' in *Journal of Australian Studies*, (43): 76–94.

Sheller, M. and J. Urry (2003) 'Mobile Transformations of "Public" and "Private" Life' in *Theory, Culture and Society*, 20(3): 107–25.

Smith, G. (1993) 'Beat Dynamics' in *National AIDS Bulletin*, 7(10): 19–21.

Smith, M.P. (2001) *Transnational Urbanism: Locating Globalization*: Oxford, Blackwell.

Smith, N. (1998) 'Giuliani Time: The Revanchist 1990s' in *Social Text*, (57): 1–20.

Smith, N. and S. Low (2006) 'Introduction: The Imperative of Public Space' in S. Low and N. Smith (Eds.) *The Politics of Public Space*: New York, Routledge.

Sorkin, M. (Ed.) (1992) *Variations on a Theme Park: The New American City and the End of Public Space*: New York, Hill and Wang.

Spearritt, P. (2000) 'The Commercialisation of Public Space' in P.N. Troy (Ed.) *Equity, Environment, Efficiency: Ethics and Economics in Urban Australia*: Melbourne, Melbourne University Press.

Staeheli, L. (1996) 'Publicity, Privacy, and Women's Political Action' in *Environment and Planning D: Society and Space*, 14(5): 601–19.

Stephenson, G. (1975) *The Design of Central Perth – Some Problems and Possible Solutions. A Study Made for the Perth Central Area Design Co-ordinating Committee*: Perth, University of Western Australian Press.

Stewart, S. (1987) 'Ceci Tuera Cela: Graffiti as Crime and Art' in J. Fekete (Ed.) *Life after Postmodernism: Essays on Value and Culture*: New York, St Martin's Press.

Strong, C. (1997) 'Contemporary Urban Art Project' in *Graffiti: Off the Wall Seminar Proceedings, University of New South Wales, 11 June*: Sydney, http://www.graffiti.nsw.gov.au.

Swivel, M. (1991) 'Public Convenience, Public Nuisance: Criminological Perspectives on "the Beat"' in *Current Issues in Criminal Justice*, 3(2): 237–49.

Tajbakhsh, K. (2001) *The Promise of the City: Space, Identity and Politics in Contemporary Social Thought*: Berkeley, University of California Press.

Taylor, C. (2004) *Modern Social Imaginaries*: Durham, Public Planet Books, Duke University Press.

Taylor, G. (1998) 'Security Management: A Sensitive Balance at Australia's Most Prominent House' in *Platypus: The Journal of the Australian Federal Police*, (60): 40–7.

Thornton, M. (1990) *The Liberal Promise – Anti-Discrimination Legislation in Australia*: Melbourne, Oxford University Press.

Thrift, N. (2004) 'Driving in the City' in *Theory, Culture and Society*, 21(4/5): 41–59.

Tilly, C. (1997) 'Parliamentarization of Popular Contention in Great Britain, 1758–1824' in *Theory and Society*, 26: 245–73.

Valentine, G. (1993) '(Hétro)sexing Space: Lesbian Perceptions and Experiences of Everyday Spaces' in *Environment and Planning D: Society and Space*, 11: 335–413.

Valentine, G. (1996) 'Children Should be Seen and Not Heard: The Production and Transgression of Adults' Public Space' in *Urban Geography*, 17(3): 205–20.

Warden, J. (1995) *A Bunyip Democracy: The Parliament and Australian National Identity*: Canberra, Australian Government Publishing Service.

Wark, M. (1990) 'Bum Rap?' in *Australian Left Review*, (120): 34–7.

Warner, M. (1990) *The Letters of the Republic: Publication and the Public Sphere in Eighteenth-Century America*: Cambridge, MA, Harvard University Press.

Warner, M. (2000) *The Trouble with Normal: Sex, Politics, and the Ethics of Queer Life*: New York, Free Press.

Warner, M. (2002) *Publics and Counterpublics*: New York, Zone.

Watson, S. (2002) 'The Public City' in J. Eade and C. Mele (Eds.) *Understanding the City: Contemporary and Future Perspectives*: Oxford, Blackwell.

Watson, S. and K. Gibson (1995) 'Postmodern Politics and Planning: A Postscript' in S. Watson and K. Gibson (Eds.) *Postmodern Cities and Spaces*: Oxford, Blackwell.

Weintraub, J. (1997) 'The Theory and Politics of the Public/Private Distinction' in J. Weintraub and K. Kumar (Eds.) *Public and Private in Thought and Practice: Perspectives on a Grand Dichotomy*: Chicago, University of Chicago Press.

Western Australian Chamber of Commerce and Industry (1992) 'Perth A Capital City for Western Australia' in Australian Institute of Urban Studies WA Branch (Ed.) *Capital City Planning for Perth, the Capital of Western Australia*: Perth, Australian Institute of Urban Studies.

Western Australian Department of Indigenous Affairs (2005) *Overcoming Indigenous Disadvantage in Western Australia*: Perth, Western Australian Department of Indigenous Affairs.

White, R. (1996) 'No-Go in the Fortress City: Young People, Inequality, and Space' in *Urban Policy and Research*, 14(1): 37–50.

Whitehead, G. (1992) 'Melbourne's Public Gardens – A Family Tree' in *Victorian Historical Journal*, 63(2–3): 101–17.

Wilson, E. (1991) *The Sphinx in the City*: London, Virago.

Wilson, J.Q. and G.L. Kelling (1982) 'Broken Windows: The Police and Neighborhood Safety' in *Atlantic Monthly*, (March): 29–32.

Wilson, P. and P. Healy (1986) *Vandalism and Graffiti on State Rail*: Canberra, Australian Institute of Criminology.

Wimsatt, W.U. (2000 (1994)) *Bomb the Suburbs*: New York, Soft Skull Press.

Wimsatt, W.U. (2000 (1999)) *No More Prisons*: New York, Soft Skull Press.

Winther, J. (1994) *Police Operations – Investigating Complaints against Persons Involved in Sexual Activity in Public Places Namely Public Toilets*: Melbourne, Victoria Police, Operations Department Support District.

Woodhead, D. (1995) ' "Surveillant Gays": HIV, Space and the Constitution of Identities' in D. Bell and G. Valentine (Eds.) *Mapping Desire: Geographies of Sexuality*: New York, Routledge.

Wotherspoon, G. (1991) *City of the Plain: History of a Gay Subculture*: Sydney, Hale and Ironmonger.

Yeoh, B.S.A. (2001) 'Postcolonial Cities' in *Progress in Human Geography*, 25(3): 456–68.

Young, I.M. (1990) *Justice and the Politics of Difference*: Princeton, Princeton University Press.

Young, I.M. (2000) *Inclusion and Democracy*: Oxford, Oxford University Press.

Youth Action and Policy Association of NSW and Youth Justice Coalition (1994) *Nobody Listens*: Sydney, Youth Action and Policy Association of NSW.

Youth Affairs Council of Western Australia (2003) 'The Public Fight for Young People to be in the City' in *Indigenous Law Bulletin*, 5(27): 8–9.

Index